STATISTICAL METHODS FOR GEOGRAPHY

STATISTICAL METHODS FOR GEOGRAPHY

PETER A. ROGERSON

SAGE Publications
London • Thousand Oaks • New Delhi

First published 2001. Reprinted 2001, 2002

SAGE Publications Ltd
6 Bonhill Street
London EC2A 4PU

SAGE Publications Inc.
2455 Teller Road
Thousand Oaks, California 91320

SAGE Publications India Pvt Ltd
32, M-Block Market
Greater Kailash - I
New Delhi 110 048

British Library Cataloguing in Publication data

A catalogue record for this book is available from the British Library

ISBN 0 7619 6287 5
ISBN 0 7619 6288 3 (pbk)

Library of Congress catalog record available

Typeset by Keyword Publishing Services Limited, UK

Printed in Great Britain by The Cromwell Press Ltd,
Trowbridge, Wiltshire

Contents

Preface

The development of geographic information systems (GIS), an increasing availability of spatial data, and recent advances in methodological techniques have all combined to make this an exciting time to study geographic problems. During the late 1970s and throughout the 1980s there had been, among many, an increasing disappointment in, and questioning of, the methods developed during the quantitative revolution of the 1950s and 1960s. Perhaps this reflected expectations that were initially too high – many had thought that sheer computing power coupled with sophisticated modeling would "solve" many of the social problems faced by urban and rural regions. But the poor performance of spatial analysis that was perceived by many was at least partly attributable to a limited capability to access, display, and analyze geographic data. During the last decade, geographic information systems have been instrumental not only in providing us with the capability to store and display information, but also in encouraging the provision of spatial datasets and the development of appropriate methods of quantitative analysis. Indeed, the GIS revolution has served to make us aware of the critical importance of spatial analysis. Geographic information systems do not realize their full potential without the ability to carry out methods of statistical and spatial analysis, and an appreciation of this dependence has helped to bring about a renaissance in the field.

Significant advances in quantitative geography have been made during the past decade, and geographers now have both the tools and the methods to make valuable contributions to fields as diverse as medicine, criminal justice, and the environment. These capabilities have been recognized by those in other fields, and geographers are now routinely called upon as members of interdisciplinary teams studying complex problems. Improvements in computer technology and computation have led quantitative geography in new directions. For example, the new field of geocomputation (see, e.g., Longley et al. 1998) lies at the intersection of computer science, geography, information science, mathematics, and statistics. The recent book by Fotheringham *et al.* (2000) also summarizes many of the new research frontiers in quantitative geography.

The purpose of this book is to provide undergraduate and beginning graduate students with the background and foundation that are necessary to be prepared for spatial analysis in this new era. I have deliberately adopted a fairly traditional approach to statistical analysis, along with several notable differences. First, I have attempted to condense much of the material found in the beginning of introductory texts on the subject. This has been done so that there is an opportunity to progress further in important areas such as regression analysis and the analysis of geographic patterns in one semester's time. Regression is by far the most common method used in geographic analysis, and it is unfortunate that it is often left to be covered hurriedly in the last week or two of a "Statistics in Geography" course.

The level of the material is aimed at upper-level undergraduate and beginning graduate students. I have attempted to structure the book so that it may be used as either a first-semester or a second-semester text. It may be used for a second-semester course by those students who already possess some background in introductory statistical concepts. The introductory material here would then serve as a review. However, the book is also meant to be fairly self-contained, and thus it should also be appropriate for those students learning about statistics in geography for the first time. First-semester students, after completing the introductory material in the first few chapters, will still be able to learn about the methods used most often by geographers by the end of a one-semester course; this is often not possible with many first-semester texts.

In writing this text, I had several goals. The first was to provide the basic material associated with the statistical methods most often used by geographers. Since a very large number of textbooks provide this basic information, I also sought to distinguish it in several ways. I have attempted to provide plenty of exercises. Some of these are to be done by hand (in the belief that it is always a good learning experience to carry out a few exercises by hand, despite what may sometimes be seen as drudgery!), and some require a computer. Although teaching the reader how to use computer software for statistical analysis is *not* one of the specific aims of this book, some guidance on the use of *SPSS for Windows 9.0* is provided. It is important that students become familiar with *some* software that is capable of statistical analysis. An important skill is the ability to sift through output and pick out what is important from what is not. Different software will produce output in different forms, and it is also important to be able to pick out relevant information whatever the arrangement of output.

In addition, I have tried to give students some appreciation of the special issues and problems raised by the use of geographic data. Straightforward application of the standard methods ignores the special nature of spatial data, and can lead to misleading results. Topics such as spatial autocorrelation and the modifiable areal unit problem are introduced to provide a good awareness of these issues, their consequences, and potential solutions. Because a full treatment of these topics would require a higher level of mathematical sophistication, they are not covered fully, but pointers to other, more advanced work and to examples are provided.

Another objective has been to provide some examples of statistical analysis that appear in the recent literature in geography. This should help to make clear the relevance and timeliness of the methods. Finally, I have attempted to point out some of the limitations of a confirmatory statistical perspective, and have directed the student to some of the newer literature on exploratory spatial data analysis. Despite the popularity and importance of exploratory methods, inferential statistical methods remain absolutely essential in the assessment of hypotheses. This text aims to provide a background in these statistical methods and to illustrate the special nature of geographic data.

A Guggenheim Fellowship afforded me the opportunity to finish the manuscript during a sabbatical leave in England. I would like to thank Paul Longley for his careful reading of an earlier draft of the book. His excellent suggestions for revision have led to a better final result. Yifei Sun and Ge Lin also provided comments that were very helpful in revising earlier drafts. Art Getis, Stewart Fotheringham, Chris Brunsdon, Martin Charlton, and Ikuho Yamada suggested changes in particular sections, and I am grateful for their assistance. Emil Boasson and my daughter, Bethany Rogerson, assisted with the production of the figures. I am thankful for the thorough job carried out by Richard Cook of Keyword in editing the manuscript. Finally, I would like to thank Robert Rojek at Sage Publications for his encouragement and guidance.

1 Introduction to Statistical Analysis in Geography

1.1 Introduction

The study of geographic phenomena often requires the application of statistical methods to produce new insight. The following questions serve to illustrate the broad variety of areas in which statistical analysis has recently been applied to geographic problems:

(1) How do blood lead levels in children vary over space? Are the levels randomly scattered throughout the city, or are there discernible geographic patterns? How are any patterns related to the characteristics of both housing and occupants? (Griffith et al. 1998).
(2) Can the geographic diffusion of democracy that has occurred during the post-World War II era be described as a steady process over time, or has it occurred in waves, or have their been "bursts" of diffusion that have taken place during short time periods? (O'Loughlin et al. 1998).
(3) What are the effects of global warming on the geographic distribution of species? For example, how will the type and spatial distribution of tree species change in particular areas? (MacDonald et al. 1998).
(4) What are the effects of different marketing strategies on product performance? For example, are mass-marketing strategies effective, despite the more distant location of their markets? (Cornish 1997).

These studies all make use of statistical analysis to arrive at their conclusions. Methods of statistical analysis play a central role in the study of geographic problems – in a survey of articles that had a geographic focus, Slocum (1990) found that 53% made use of at least one mainstream quantitative method. The role of statistical analysis in geography may be placed within a broader context through its connection to the "scientific method," which provides a more general framework for the study of geographic problems.

1.2 The Scientific Method

Social scientists as well as physical scientists often make use of the *scientific method* in their attempts to learn about the world. Figure 1.1 illustrates this

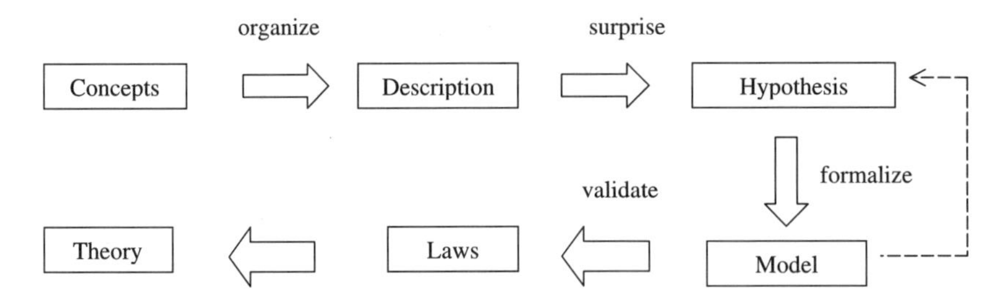

Figure 1.1 **The scientific method**

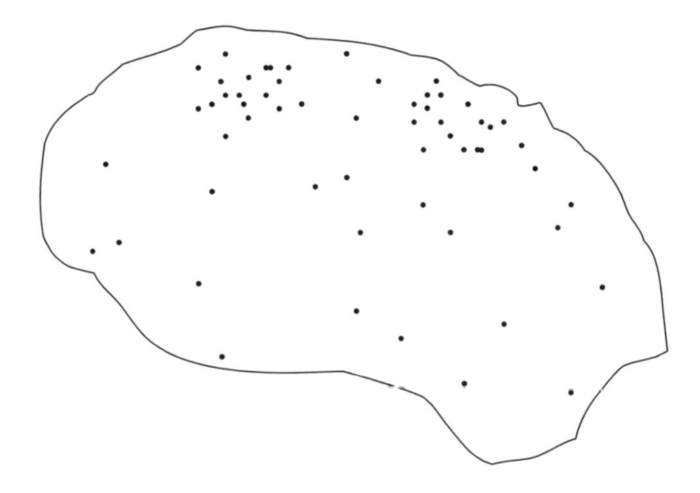

Figure 1.2 **Distribution of cancer cases**

method, from the initial attempts to organize ideas about a subject to the building of a theory.

Suppose that we are interested in describing and explaining the spatial pattern of cancer cases in a metropolitan area. We might begin by plotting recent incidences on a map. Such descriptive exercises often lead to an unexpected result – in Figure 1.2, we perceive two fairly distinct clusters of cases. The surprising results generated through the process of description naturally lead us to the next step on the route to explanation by forcing us to generate hypotheses about the underlying process. A "rigorous" definition of the term *hypothesis* is a proposition whose truth or falsity is capable of being tested. Though in the social sciences we do not always expect to come to firm conclusions in the form of "laws," we can also think of hypotheses as potential answers to our initial surprise. For example, one hypothesis in the present example is that the pattern of cancer cases is related to the distance from local power plants.

To test the hypothesis, we need a *model*, which is a device for simplifying reality so that the relationship between variables may be more clearly studied.

Whereas a hypothesis might suggest a relationship between two variables, a model is more detailed, in the sense that it suggests the nature of the relationship between the variables. In our example, we might speculate that the likelihood of cancer declines as the distance from a power plant increases. To test this model, we could plot cancer rates for a subarea versus the distance the subarea centroid was from a power plant. If we observe a downward sloping curve, we have gathered some support for our hypothesis (see Figure 1.3).

Models are validated by comparing observed data with what is expected. If the model is a good representation of reality, there will be a close match between the two. If observations and expectations are far apart, we need to "go back to the drawing board" and come up with a new hypothesis. It might be the case, for example, that the pattern in Figure 1.2 is due simply to the fact that the population itself is clustered. If this new hypothesis is true, or if there is evidence in favor of it, the spatial pattern of cancer then becomes understandable; a similar rate throughout the population generates apparent cancer clusters because of the spatial distribution of the population.

Though a model is often used to learn about a particular situation, more often one also wishes to learn about the underlying process that led to it. We would like to be able to *generalize* from one study to statements about other situations. One reason for studying the spatial pattern of cancer cases is to determine whether there is a relationship between cancer rates and the distance to *specific* power plants; a more general objective is to learn about the relationship between cancer rates and the distance to *any* power plant. One way of making such generalizations is to accumulate a lot of evidence. If we were to repeat our analysis in many locations throughout a country, and if our findings were similar in all cases, we would have uncovered an empirical generalization. In a strict sense, *laws* are sometimes defined as universal

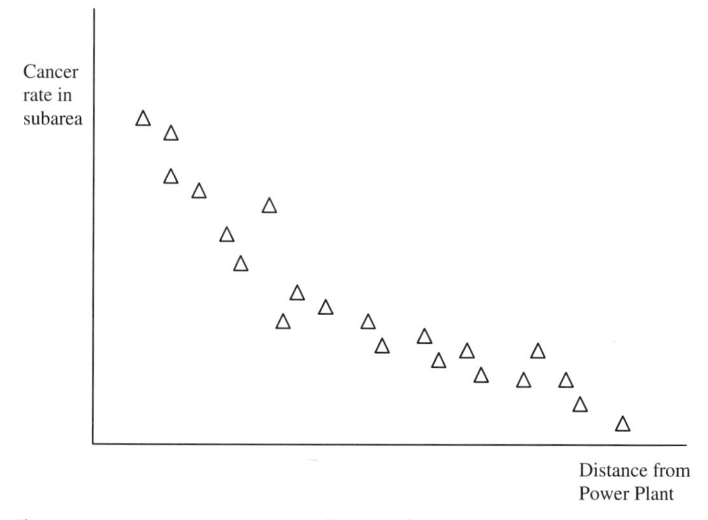

Figure 1.3 **Cancer rates versus distance from power plant**

statements of unrestricted range. In our example, our generalization would not have unrestricted range, and we might want, for example, to confine our generalization or empirical law to power plants and cancer cases in a particular country.

Einstein called theories "free creations of the human mind." In the context of our diagram, we may think of theories as collections of generalizations or laws. The whole collection is greater than the sum of its parts in the sense that it gives greater insight than that produced by the generalizations or laws alone. If for example, we generate other empirical laws that relate cancer rates to other factors, such as diet, we begin to build a theory of the spatial variation in cancer rates.

Statistical methods occupy a central role in the scientific method, as portrayed in Figure 1.1, because they allow us to suggest and test hypotheses using models. In the following section, we will review some of the important types of statistical approaches in geography.

1.3 Exploratory and Confirmatory Approaches in Geography

The scientific method provides us with a structured approach to answering questions of interest. At the core of the method is the desire to form and test *hypotheses*. As we have seen, hypotheses may be thought of loosely as potential answers to questions. For instance, a map of snowfall may suggest the hypothesis that the distance away from a nearby lake may play an important role in the distribution of snowfall amounts.

Geographers use spatial analysis within the context of the scientific method in at least two distinct ways. *Exploratory* methods of analysis are used to *suggest* hypotheses; *confirmatory* methods are, as the name suggests, used to help confirm hypotheses. A method of visualization or description that led to the discovery of clusters in Figure 1.2 would be an exploratory method, whereas a statistical method that confirmed that such an arrangement of points would have been unlikely to occur by chance would be a confirmatory method. In this book we will focus primarily upon confirmatory methods.

We should note here two important points. First, confirmatory methods do not always confirm or refute hypotheses – the world is too complicated a place, and the methods often have important limitations that prevent such confirmation and refutation. Nevertheless, they are important in structuring our thinking and in taking a rigorous and scientific approach to answering questions. Second, the use of exploratory methods over the past few years has been increasing rapidly. This has come about as a combination of the availability of large databases and sophisticated software (including GIS), and a recognition that confirmatory statistical methods are appropriate in some situations and not others. Throughout the book we will keep the reader aware of these points by pointing out some of the limitations of confirmatory analysis.

1.4 Descriptive and Inferential Methods

A key characteristic of geographic data that brings about the need for statistical analysis is that they may often be regarded as a sample from a larger population. *Descriptive* statistical analysis refers to the use of particular methods that are used to describe and summarize the characteristics of the sample, whereas *inferential* statistical analysis refers to the methods that are used to infer something about the population from the sample. Descriptive methods fall within the class of exploratory techniques; inferential statistics lie within the class of confirmatory methods.

1.4.1 Overview of Descriptive Analysis

Suppose that we wish to learn something about the commuting behavior of residents in a community. Perhaps we are on a committee that is investigating the potential implementation of a public transit alternative, and we need to know how many minutes, on average, it takes people to get to work by car. We do not have the resources to ask everyone, and so we decide to take a sample of automobile commuters. Let's say we survey $n = 30$ residents, asking them to record their average time it takes to get to work. We receive the responses shown in panel (a) of Table 1.1.

We begin our descriptive analysis by summarizing the information. The *sample mean* commuting time is simply the average of our observations; it is found by adding all of the individual responses and dividing by thirty.

Table 1.1 **Commuting data**

(a) *Data on individuals*

Individual no.	Commuting time (min.)	Individual no.	Commuting time (min.)
1	5	16	42
2	12	17	31
3	14	18	31
4	21	19	26
5	22	20	24
6	36	21	11
7	21	22	19
8	6	23	9
9	77	24	44
10	12	25	21
11	21	26	17
12	16	27	26
13	10	28	21
14	5	29	24
15	11	30	23

(b) *Ranked commuting times*

5, 5, 6, 9, 10, 11 , 11, 12, 12, 14, 16, 17, 19, 21, 21, 21, 21, 21, 22, 23, 24, 24, 26, 26, 31, 31, 36, 42, 44, 77

The sample mean is traditionally denoted by $\bar{x}$; in our example we have $\bar{x} = 21.93$ minutes. In practice, this could sensibly be rounded to 22 minutes. The *median* time is defined as the time that splits the ranked list of commuting times in half – half of all respondents have commutes that are longer than the median, and half have commutes that are shorter. When the number of observations is odd, the median is simply equal to the middle value on a list of the observations, ranked from shortest commute to longest commute. When the number of observations is even, as it is here, we take the median to be the average of the two values in the middle of the ranked list. When the responses are ranked as in panel (b) of Table 1.1, the two in the middle are 21 and 21. The median in this case is equal to 21 minutes. The *mode* is defined as the most frequently occurring value; here the mode is also 21 minutes, since it occurs more frequently (four times) than any other outcome.

We may also summarize the data by characterizing its variability. The data range from a low of five minutes to a high of 77 minutes. The *range* is the difference between the two values – here it is equal to $77 - 5 = 72$ minutes.

The *interquartile range* is the difference between the 25th and 75th percentiles. With n observations, the 25th percentile is represented by observation $(n+1)/4$, when the data have been ranked from lowest to highest. The 75th percentile is represented by observation $3(n+1)/4$. These will often not be integers, and interpolation is used, just as it is for the median when there is an even number of observations. For the commuting data, the 25th percentile is represented by observation $(30+1)/4 = 7.75$. Interpolation between the 7th and 8th lowest observations requires that we go 3/4 of the way from the 7th lowest observation (which is 11) to the 8th lowest observation (which is 12). This implies that the 25th percentile is 11.75. Similarly, the 75th percentile is represented by observation $3(30+1)/4 = 23.25$. Since both the 23rd and 24th observations are equal to 26, the 75th percentile is equal to 26. The interquartile range is the difference between these two percentiles, or $26 - 11.75 = 14.25$.

The *sample variance* of the data (denoted s^2) may be thought of as the average squared deviation of the observations from the mean. To ensure that the sample variance gives an unbiased estimate of the true, unknown variance of the population from which the sample was drawn (denoted σ^2), s^2 is computed by taking the sum of the squared deviations, and then dividing by $n - 1$, instead of by n. Here the term *unbiased* implies that if we were to repeat this sampling many times, we would find that the average or mean of our many sample variances would be equal to the true variance. Thus the sample variance is found from

$$s^2 = \frac{\sum_{i=1}^{n} (x_i - \bar{x})^2}{n - 1} \tag{1.1}$$

where the Greek letter Σ means that we are to sum the squared deviations of the observations from the mean (notation is discussed in more detail in Chapter 2).

In our example, $s^2 = 208.13$. The *sample standard deviation* is equal to the square root of the sample variance; here we have $s = \sqrt{208.13} = 14.43$. Since the sample variance characterizes the average *squared* deviation from the mean, by taking the square root and using the standard deviation, we are putting the measure of variability back on a scale closer to that used for the mean and the original data. It is not quite correct to say that the standard deviation is the average absolute deviation of an observation from the mean, but it is close to being correct.

Since data come from distributions with different means and different degrees of variability, it is common to standardize observations. One way to do this is to transform each observation into a *z-score* by first subtracting the mean of all observations and then dividing the result by the standard deviation:

$$z = \frac{x - \bar{x}}{s} \tag{1.2}$$

z-scores may be interpreted as the number of standard deviations an observation is away from the mean. For example, the z-score for individual 1 is $(5 - 21.93)/14.3 = -1.17$. This individual has a commuting time that is 1.17 standard deviations below the mean.

We may also summarize our data by constructing *histograms*, which are vertical bar graphs. To construct a histogram, the data are first grouped into categories. The histogram contains one vertical bar for each category. The height of the bar represents the number of observations in the category (i.e., the frequency), and it is common to note the midpoint of the category on the horizontal axis. Figure 1.4 is a histogram for the commuting data, produced by *SPSS for Windows 9.0*.

Skewness measures the degree of asymmetry exhibited by the data. Figure 1.4 reveals that there are more observations below the mean than above it – this

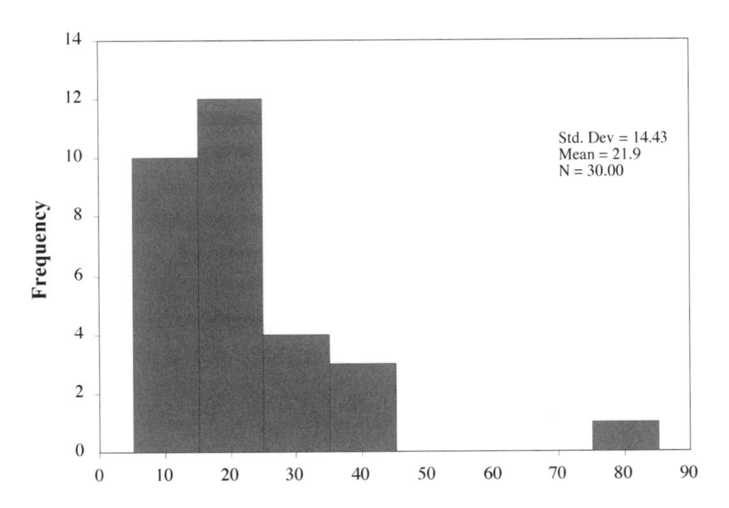

Figure 1.4 **Histogram for commuting data**

is known as positive skewness. Positive skewness can also be detected by comparing the mean and median. When the mean is greater than the median, as it is here, the distribution is positively skewed. In contrast, when there are a small number of low observations and a large number of high ones, the data exhibit negative skewness. Skewness is computed by first adding together the cubed deviations from the mean and then dividing by the product of the cubed standard deviation and the number of observations:

$$\text{skewness} = \frac{\sum_{i=1}^{n} (x_i - \bar{x})^3}{ns^3} \tag{1.3}$$

The 30 commuting times have a positive skewness of 2.06. If skewness equals zero, the histogram is symmetric about the mean.

Kurtosis measures how peaked the histogram is. Its definition is similar to that for skewness, with the exception that the fourth power is used instead of the third:

$$\text{kurtosis} = \frac{\sum_{i=1}^{n} (x_i - \bar{x})^4}{ns^4} \tag{1.4}$$

Data with a high degree of peakedness are said to be *leptokurtic*, and have values of kurtosis over 3.0. Flat histograms are *platykurtic*, and have kurtosis values less than 3.0. The kurtosis of the commuting times is equal to 6.43, and hence the distribution is relatively peaked.

Data may also be summarized via *box plots*. Figure 1.5 depicts a box plot for the commuting data. The horizontal line running through the rectangle denotes

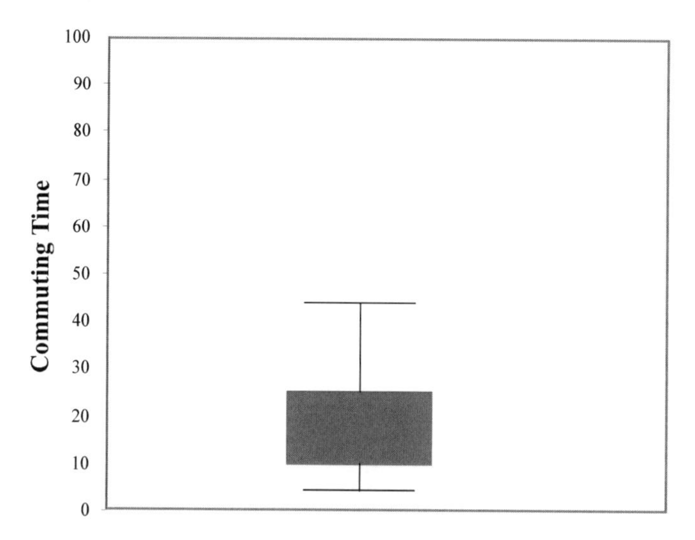

Figure 1.5 **Boxplot for commuting data**

the median (21), and the lower and upper ends of the rectangle (sometimes called the "hinges") represent the 25th and 75th percentiles, respectively. Velleman and Hoaglin (1981) note that there are two common ways to draw the "whiskers" which extend upward and downward from the hinges. One way is to send the whiskers out to the minimum and maximum values. In this case, the boxplot represents a graphical summary of what is sometimes called a "five-number summary" of the distribution (the minimum, maximum, 25th and 75th percentiles, and the median).

There are often extreme outliers in the data that are far from the mean, and in this case it is not preferable to send whiskers out to these extreme values. Instead, whiskers are sent out to the outermost observations that are still within 1.5 times the interquartile range of the hinge. All other observations beyond this are considered outliers, and are shown individually. In the commuting data, 1.5 times the interquartile range is equal to $1.5(14.25) = 21.375$. The whisker extending downward from the lower hinge extends to the minimum value of 5, since this is greater than the lower hinge (11.75) minus 21.375. The whisker extending upward from the upper hinge stops at 44, which is the highest observation less than 47.375 (which in turn is equal to the upper hinge (26) plus 21.375). Note that there is a single outlier – observation 9 – which has a value of 77 minutes.

A *stem-and-leaf* plot is an alternative way to show how common observations are. It is similar to a histogram tilted onto its side, with the actual digits of each observation's value used in place of bars. The leading digits constitute the "stem," and the trailing digits make up the "leaf." Each stem has one or more leaves, with each leaf corresponding to an observation. The visual depiction of the frequency of leaves conveys to the reader an impression of the frequency of observations that fall within given ranges. John Tukey, the designer of the stem-and-leaf plot, has said "If we are going to make a mark, it may as well be a meaningful one. The simplest – and most useful – meaningful mark is a digit." (Tukey 1972, p. 269). For the commuting data, which have at most two-digit values, the first digit is the "stem," and the second is the "leaf" (see Figure 1.6).

1.4.2 Overview of Inferential Analysis

Since we did not interview everyone, we do not know the true mean commuting time (which we denote μ) that characterizes the entire community. (Note that we use regular, Roman letters to indicate sample means and variances, and that we use Greek letters to represent the corresponding, unknown population values. This is a common notational convention that we will use throughout.) We have an estimate of the true mean from our sample mean, but it is also desirable to make some sort of inferential statement about μ that quantifies our uncertainty regarding the true mean. Clearly we would be less uncertain about the true mean if we had taken a larger

```
Frequency      Stem &  Leaf

     .00          0 .
    4.00          0 .  5569
    6.00          1 .  011224
    3.00          1 .  679
    9.00          2 .  111112344
    2.00          2 .  66
    2.00          3 .  11
    1.00          3 .  6
    2.00          4 .  24
    1.00 Extremes       (>=77)

Stem width:        10.00
Each leaf:          1 case(s)
```

Figure 1.6 **Stem-and-leaf plot for commuting data**

sample, and we would also be less uncertain about the true mean if we knew there was less variability in the population values (that is, if σ^2 were lower). Although we don't know the "true" variance of commuting times (σ^2), we do have an estimate of it (s^2).

In the next chapter, we will learn how to make inferences about the population mean from the sample mean. In particular we will learn how to test hypotheses regarding the mean (e.g., could the "true" commuting time in our population be equal to $\mu = 30$ minutes?), and we will also learn how to place confidence limits around the mean to make statements such as "we are 95% confident that the true mean lies ± 3.5 minutes from the observed mean."

To illustrate some common inferential questions using another example, suppose you are handed a coin, and you are asked to determine whether it is a "fair" one (that is, the likelihood of a "head" is the same as the likelihood of a "tail"). One natural way to gather some information would be to flip the coin a number of times. Suppose you flip the coin ten times, and you observe heads eight times. An example of a descriptive statistic is the observed proportion of heads – in this case $8/10 = 0.8$. We enter the realm of inferential statistics when we attempt to pass judgement on whether the coin is "fair". We plan to do this by *inferring* whether the coin is fair, on the basis of our sample results. Eight heads is more than the four, five, or six that might have made us more comfortable in a declaration that the coin is fair, but is eight heads really enough to say that the coin is *not* a fair one?

There are at least two ways to go about answering the question of whether the coin is a fair one. One is to ask what *would* happen if the coin *were* fair, and to simulate a series of experiments identical to the one just carried out. That is, if we could repeatedly flip a known fair coin ten times, each time recording the number of heads, we would learn just how unusual a total of eight heads actually was. If eight heads comes up quite frequently with the fair coin, we will judge our original coin to be fair. On the other hand, if eight heads is an extremely rare event for a fair coin, we will conclude that our original coin is not fair.

To pursue this idea, suppose you arrange to carry out such an experiment 100 times. For example, one might have 100 students in a large class each flip a coin that is known to be fair ten times. Upon pooling together the results, suppose you find the results shown in Table 1.2. We see that eight heads occurred 8% of the time.

We still need a guideline to tell us whether our observed outcome of eight heads should lead us to the conclusion that the coin is (or is not) fair. The usual guideline is to ask how likely a result equal to or more extreme than the observed one is, *if* our initial, baseline hypothesis that we possess a fair coin (called the *null* hypothesis) is true. A common practice is to accept the null hypothesis if the likelihood of a result more extreme than the one we observed is more than 5%. Hence we would accept the null hypothesis of a fair coin if our experiment showed that eight or more heads was not uncommon and in fact tended to occur more than 5% of the time.

Alternatively, we wish to reject the null hypothesis that our original coin is a fair one if the results of our experiment indicate that eight or more heads out of ten is an uncommon event for fair coins. If fair coins give rise to eight or more heads less than 5% of the time, we decide to reject the null hypothesis and conclude that our coin is not fair.

In the example above, eight or more heads occurred 12 times out of 100, when a fair coin was flipped ten times. The fact that events as extreme as, or more extreme than the one we observed will happen 12% of the time with a *fair* coin leads us to accept the inference that our original coin is a fair one. Had we observed nine heads with our original coin, we would have judged it to be unfair, since events as rare or more rare than this (namely where the number of heads is equal to 9 or 10) occurred only four times in the one hundred trials of a fair coin. Note, too, that our observed result does not prove that the coin *is* unbiased. It still *could* be unfair; there is, however, insufficient evidence to support the allegation.

Table 1.2 **Hypothetical outcome of 100 experiments of ten coin tosses each**

No. of heads	Frequency of occurrence
0	0
1	1
2	4
3	8
4	15
5	22
6	30
7	8
8	8
9	3
10	1

The approach just described is an example of the Monte Carlo method, and several examples of its use are given in Chapter 8. A second way to answer the inferential problem is to make use of the fact that this is a *binomial* experiment; in Chapter 2 we will learn how to use this approach.

1.5 The Nature of Statistical Thinking

The American Statistical Association (1993, cited in Mallows 1998) notes that statistical thinking is

(a) the appreciation of uncertainty and data variability, and their impact on decision making; and
(b) the use of the scientific method in approaching issues and problems.

Mallows (1998), in his Presidential Address to the American Statistical Association, argues that statistical thinking is not simply common sense, nor is it simply the scientific method. Rather, he suggests that statisticians give more attention to questions that arise in the beginning of the study of a problem or issue. In particular, Mallows argues that statisticians should (a) consider what data are relevant to the problem, (b) consider how relevant data can be obtained, (c) explain the basis for all assumptions, (d) lay out the arguments on all sides of the issue, and only then (e) formulate questions that can be addressed by statistical methods. He feels that too often statisticians rely too heavily on (e), as well as on the actual use of the methods that follow. His ideas serve to remind us that statistical analysis is a comprehensive exercise – it does not consist of simply "plugging numbers into a formula" and reporting a result. Instead, it requires a comprehensive assessment of questions, alternative perspectives, data, assumptions, analysis, and interpretation.

Mallows defines statistical thinking as that which "concerns the relation of quantitative data to a real-world problem, often in the presence of uncertainty and variability. It attempts to make precise and explicit what the data has to say about the problem of interest." Throughout the remainder of this book, we will learn how various methods are used and implemented, but we will also learn how to interpret the results and understand their limitations. Too often students working on geographic problems have only a sense that they "need statistics," and their response is to seek out an expert on statistics for advice on how to get started. The statistician's first reply should be in the form of questions: (1) What is the problem? (2) What data do you have, and what are its limitations? (3) Is statistical analysis relevant, or is some other method of analysis more appropriate? It is important for the student to think first about these questions. Perhaps simple description will suffice to achieve the objective. Perhaps some sophisticated inferential analysis will be necessary. But the subsequent course of events should be driven by the substantive problems and questions of interest, as constrained by data availability and quality. It should not be driven by a feeling that one needs to use statistical analysis simply for the sake of doing so.

1.6 Some Special Considerations with Spatial Data

Fotheringham and Rogerson (1993) categorize and discuss a number of general issues and characteristics associated with problems in spatial analysis. It is essential that those working with spatial data have an awareness of these issues. Although all of their categories are relevant to spatial *statistical* analysis, among those that are most pertinent are:

(a) the modifiable areal unit problem;
(b) boundary problems;
(c) spatial sampling procedures;
(d) spatial autocorrelation.

1.6.1 Modifiable Areal Unit Problem

The modifiable areal unit problem refers to the fact that results of statistical analyses are sensitive to the zoning system used to report aggregated data. Many spatial datasets are aggregated into zones, and the nature of the zonal configuration can influence interpretation quite strongly. Panel (a) of Figure 1.7 shows one zoning system and panel (b) another. The arrows represent migration flows. In panel (a) no interzonal migration is reported, whereas an interpretation of panel (b) would lead to the conclusion that there was a strong southward movement. More generally, many of the statistical tools described in the following chapters would produce different results had different zoning systems been in effect.

The modifiable areal unit problem has two different aspects that should be appreciated. The first is related to the placement of zonal boundaries, for zones or subregions of a given size. If we were measuring mobility rates, we could overlay a grid of square cells on the study area. There are many different ways that the grid could be placed, rotated, and oriented on the study area. The second aspect has to do with geographic scale. If we were to replace the grid with another grid of larger square cells, the results of the analysis would be different. Migrants, for example, are less likely to cross cells in the larger grid than they are in the smaller grid.

As Fotheringham and Rogerson (1993) note, GIS technology now facilitates the analysis of data using alternative zoning systems, and it should become more routine to examine the sensitivity of results to modifiable areal units.

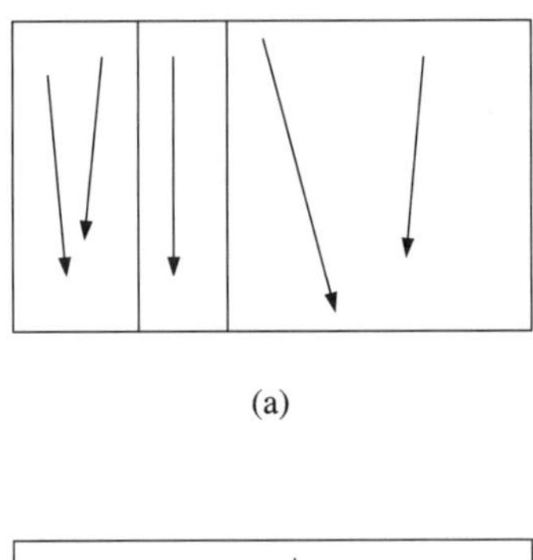

(a)

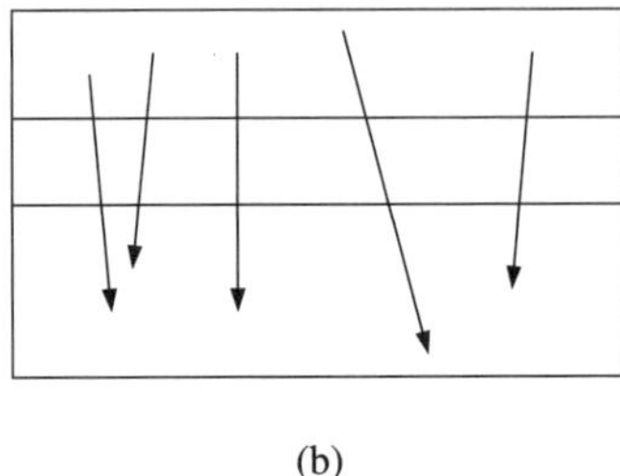

(b)

Figure 1.7 **Two alternative zoning systems for migration data. Arrows show origins and destinations of migrants**

1.6.2 Boundary Problems

Study areas are bounded, and it is important to recognize that events just outside the study area can affect those inside it. If we are investigating the market areas of shopping malls in a county, it would be a mistake to neglect the influence of a large mall located just outside the county boundary. One solution is to create a buffer zone around the area of study to include features that affect analysis within the primary area of interest. An example of the use of buffer zones in point pattern analysis is given in Chapter 8.

Both the size and shape of areas can affect measurement and interpretation. There are many migrants leaving Rhode Island each year, but this is partially due to the state's small size – almost any move will be a move out of the state! Similarly, Tennessee experiences more out-migration than other states with the same land area, in part because of its narrow rectangular shape. This is because individuals in Tennessee live, on average, closer to the border than do individuals in other states with the same area. A move of given length in some random direction is therefore more likely to take the Tennessean outside of the state.

1.6.3 Spatial Sampling Procedures

Statistical analysis is based upon sample data. Usually one assumes that sample observations are taken randomly from some larger population of interest. If we are interested in sampling point locations to collect data on vegetation or soil, for example, there are many ways to do this. One could choose x- and y-coordinates randomly; this is known as a *simple random sample*. Another alternative would be to choose a *stratified* spatial sample, making sure that we chose a predetermined number of observations from each of several subregions, with simple random sampling within subregions. Alternative methods of sampling are discussed in more detail in Section 3.7.

1.6.4 Spatial Autocorrelation

Spatial autocorrelation refers to the fact that the value of a variable at one point in space is related to the value of that same variable in a nearby location. The travel behavior of residents in a household is likely to be related to the travel behavior of residents in nearby households, because both households have similar accessibility to other locations. Hence observations of the two households are not likely to be independent, despite the requirement of statistical independence for standard statistical analysis. Spatial autocorrelation can therefore have serious effects on statistical analyses, and hence lead to misinterpretation. It is treated in more detail in Chapter 8.

1.7 Descriptive Statistics in *SPSS for Windows 9.0*

1.7.1 Data Input

After starting *SPSS*, data are input for the variable or variables of interest. Each column represents a variable. For the commuting example set out in Table 1.1, the thirty observations were entered into the first column of the spreadsheet. Alternatively, respondent ID could have been entered into the first column (i.e., the sequence of integers, from 1 to 30), and the commuting times would then have been entered in the second column). The order that the data are entered into a column is unimportant.

1.7.2 Descriptive Analysis

Simple descriptive statistics. Once the data are entered, click on Analyze (or Statistics, in older versions of *SPSS for Windows*). Then click on Descriptive Statistics. Then click on Explore. A split box will appear on the screen; move the variable or variables of interest from the left box to the box on the right that is headed "Dependent List" by highlighting the variable(s) and clicking on the arrow. Then click on OK.

Table 1.3 ***SPSS* output for data of Table 1.1**

Descriptives

			Statistic	Std. Error
VAR00001	Mean		21.9333	2.6340
	95% Confidence Interval for Mean	Lower Bound	16.5463	
		Upper Bound	27.3204	
	5% Trimmed Mean		20.4259	
	Median		21.0000	
	Variance		208.133	
	Std. Deviation		14.4268	
	Minimum		5.00	
	Maximum		77.00	
	Range		72.00	
	Interquartile Range		14.2500	
	Skewness		2.057	.427
	Kurtosis		6.434	.833

Other options. Options for producing other related statistics and graphs are available. To produce a histogram for instance, before clicking OK above, click on Plots, and you can then check a box to produce a histogram. Then click on Continue and OK.

Results. Table 1.3 displays results of the output. In addition to this table, boxplots (Figure 1.5), stem and leaf displays (Figure 1.6) and, optionally, histograms (Figure 1.4) are also produced.

Exercises

1. The 236 values that appear below are the 1990 median household incomes (in dollars) for the 236 census tracts of Buffalo, New York.

(a) For the first 19 tracts, find the mean, median, range, interquartile range, standard deviation, variance, skewness, and kurtosis using only a calculator (though you may want to check your results using a statistical software package). In addition, construct a stem-and-leaf plot, a box plot, and a histogram for these 19 observations.

(b) Use a statistical software package to repeat part (a), this time using all 236 observations.

(c) Comment on your results. In particular, what does it mean to find the mean of a set of medians? How do the observations that have a value of 0 affect the results? Should they be included? How might the results differ if a different geographic scale were chosen?

22342, 19919, 8187, 15875, 17994, 30765, 31347, 27282, 29310, 23720, 22033, 11706, 15625, 6173, 15694, 7924, 10433, 13274, 17803, 20583, 21897, 14531, 19048, 19850, 19734, 18205, 13984, 8738, 10299, 10678, 8685, 13455,

14821, 23722, 8740, 12325, 10717, 21447, 11250, 16016, 11509, 11395, 19721, 23231, 21293, 24375, 19510, 14926, 22490, 21383, 25060, 22664, 8671, 31566, 26931, 0, 24965, 34656, 24493, 21764, 25843, 32708, 22188, 19909, 33675, 15608, 15857, 18649, 21880, 17250, 16569, 14991, 0, 8643, 22801, 39708, 17096, 20647, 30712, 19304, 24116, 17500, 19106, 17517, 12525, 13936, 7495, 10232, 6891, 16888, 42274, 43033, 43500, 22257, 22931, 31918, 29072, 31948, 36229, 33860, 32586, 32606, 31453, 32939, 30072, 32185, 35664, 27578, 23861, 18374, 26563, 30726, 33614, 30373, 28347, 37786, 48987, 56318, 49641, 85742, 43229, 53116, 44335, 30184, 36744, 39698, 0, 21987, 66358, 46587, 26934, 27292, 31558, 36944, 43750, 49408, 37354, 31010, 35709, 32913, 25594, 25612, 28980, 28800, 28634, 18958, 26515, 24779, 21667, 24660, 29375, 29063, 30996, 45645, 39312, 34287, 35533, 27647, 24342, 22402, 28967, 39083, 28649, 23881, 31071, 27412, 27943, 34500, 19792, 41447, 35833, 41957, 14333, 12778, 20000, 19656, 22302, 33475, 26580, 0, 24588, 31496, 30179, 33694, 36193, 41921, 35819, 39304, 38844, 37443, 47873, 41410, 34186, 36798, 38508, 38382, 37029, 48472, 38837, 40548, 35165, 39404, 34281, 24615, 34904, 21964, 42617, 58682, 41875, 40370, 24511, 31008, 16250, 29600, 38205, 35536, 35386, 36250, 31341, 33790, 31987, 42113, 37500, 33841, 37877, 35650, 28556, 27048, 27736, 30269, 32699, 28988, 22083, 27446, 76306, 19333

2. Ten migration distances corresponding to the distances moved by recent migrants are observed (in miles): 43, 6, 7, 11, 122, 41, 21, 17, 1, 3. Find the mean and standard deviation, and then convert all observations into z-scores.

3. The probability of commuting by train in a community is 0.1. A survey of residents in a particular neighborhood finds that four out of ten commute by train. We wish to conclude either that (a) the "true" commuting rate in the neighborhood is 0.1, and we have just witnessed four out of ten as a result of sampling fluctuation, or (b) the "true" commuting rate in the neighborhood is greater than 0.1, and it is very unlikely that we would have observed four out of ten train commuters if the true rate was 0.1.

Decide which choice is best via the following steps, using the random number table in Table A.1 of Appendix A:

(1) take a series of ten random digits, and then count and record the number of "0"s; these will represent the number of train commuters in a sample of ten, where the "true" commuting probability is 0.1.
(2) Repeat step 1 twenty times.
(3) Arrive at either conclusion (a) or (b). You should arrive at conclusion (b) if you had four or more commuters either once, or not at all, in the twenty repetitions (since one out of twenty is equal to 0.05, or 5%).

2 Probability and Probability Models

LEARNING OBJECTIVES

- Review of mathematical notation and ordering of mathematical operations
- Introduction to probability concepts, including (a) sample spaces as potential outcomes of experiments, (b) assignment of probabilities to individual outcomes
- Binomial and normal distributions
- Confidence intervals for the sample mean
- Examples of applications based upon simple probability models

In Chapter 1, we had our first glimpse into some of the concepts that are used both to describe sample data and to make inferences. In this chapter, we will build upon these concepts. After reviewing mathematical conventions and notation in the beginning of the chapter, we will explore some of the basic concepts of probability, which form the basis for statistical inference.

2.1 Mathematical Conventions and Notation

The amount of mathematical notation used in this book is actually quite small, but, nevertheless, it is useful to review some basic notation and mathematical conventions.

2.1.1 Mathematical Conventions

By the term "mathematical conventions" we are not referring here to the gatherings of mathematicians at conferences, but rather to the standards that are used in the writing and use of mathematical material. The primary conventions we are concerned with are those regarding parentheses and the ordering of mathematical operations. In a mathematical expression, one performs operations in the following order, arranged from operations performed first to those performed last:

(1) Factorials (the factorial of an integer m is the product of the integers from 1 to m, and is further defined below).
(2) Powers and roots.

(3) Multiplication and division.
(4) Addition and subtraction.

Thus the expression

$$3 + 10/5^2 \tag{2.1}$$

is evaluated by first squaring 5, then finding 10/25 = 0.4, and then adding 3 to find the result of 3.4. One does not simply go from left to right; if you did, you would incorrectly add 10 to 3, then divide by 5 to get 2.6, and then square 2.6 for a final (incorrect) answer of 6.76.

If there is more than one operation in any of the four categories above, one carries out those particular operations from left to right. Thus, to evaluate

$$3 + 10/5 + 6 * 7 \tag{2.2}$$

one would do the division first and the multiplication second, yielding

$$3 + 2 + 42 = 47 \tag{2.3}$$

Although it would be unusual to see it written this way,

$$6/3/3 \tag{2.4}$$

is equal to 2/3, since 6/3 would be carried out first.

Operations within parentheses are always performed before those that are not within parentheses, and those within nested parentheses are dealt with by performing the operations within the innermost set of parentheses first. So, for example,

$$3 * ((5 + 3)^2/2) + 4 = 3 * (8^2/2) + 4 = 3 * 32 + 4 = 100 \tag{2.5}$$

Although these basic principles are taught before the high-school years, it is not uncommon to need a little review! It is important to realize too that it is not just students of statistics that need brushing up – software developers and decision-makers sometimes do not abide by these conventions. For example, new variables that are created within the geographic information system (GIS) *ArcView* 3.1 are created by simply carrying out operations from left to right. Although parentheses are recognized, the fundamental order of operations, as outlined above, is not! This leads to visions of planners and others all over the world making decisions based upon inaccurate information!

Suppose we have data on the proportion of people commuting by train (variable 1), the number of people who commute by bus (variable 2), and the total number of commuters (variable 3) for a number of census tracts in our database. Thinking that *ArcView* will surely use the standard order of mathematical operations, we compute a new variable reflecting the proportion of people who commute by bus or train (variable 4) via

$$\text{Var. } 4 = \text{Var. } 1 + \text{Var. } 2/\text{Var. } 3 \tag{2.6}$$

ArcView will provide us with a column of answers where

$$\text{Var. } 4 = (\text{Var. } 1 + \text{Var. } 2)/\text{Var. } 3 \tag{2.7}$$

when in fact what we wanted was

$$\text{Var. } 4 = \text{Var. } 1 + (\text{Var. } 2/\text{Var. } 3) \tag{2.8}$$

One way of ensuring that problems like this do not arise is to use extra sets of parentheses, as in the last equation (and, in fact, to obtain the desired variable within ArcView, they *must* be used).

2.1.2 Mathematical Notation

The mathematical notation used most often in this book is the summation notation. The Greek letter Σ is used as a shorthand way of indicating that a sum is to be taken. For example,

$$\sum_{i=1}^{i=n} x_i \tag{2.9}$$

denotes that the sum of n observations is to be taken; the expression is equivalent to

$$x_1 + x_2 + \cdots + x_n \tag{2.10}$$

The "i=1" under the symbol refers to where the sum of terms begins, and the "i=n" refers to where it terminates. Thus

$$\sum_{i=3}^{i=5} x_i = x_3 + x_4 + x_5 \tag{2.11}$$

implies that we are to sum only the third, fourth, and fifth observations. There are a number of rules that govern the use of this notation. These may be summarized as follows, where a is a constant, n is the number of observations, and x and y are variables:

$$\left.\begin{aligned} &\sum_{i=1}^{i=n} a = na \\ &\sum_{i=1}^{i=n} ax_i = a\sum_{i=1}^{i=n} x_i \\ &\sum_{i=1}^{i=n} (x_i + y_i) = \sum_{i=1}^{i=n} x_i + \sum_{i=1}^{i=n} y_i \end{aligned}\right\} \tag{2.12}$$

The first states that summing a constant n times yields a result of an. Thus

$$\sum_{i=1}^{i=3} 4 = 4 + 4 + 4 = 4 * 3 = 12 \tag{2.13}$$

The second rule in (2.12) indicates that constants may be taken outside of the summation sign. So, for example,

$$\sum_{i=1}^{i=3} 3x_i = 3\sum_{i=1}^{i=3} x_i = 3(x_1 + x_2 + x_3) \tag{2.14}$$

The third rule implies that the order of addition does not matter when sums of sums are being taken.

Other conventions include

$$\left.\begin{aligned} &\sum_{i=1}^{i=n} x_i y_i = x_1 y_1 + x_2 y_2 + \cdots + x_n y_n \\ &\sum_{i=1}^{i=n} x_i^2 = x_1^2 + x_2^2 + \cdots + x_n^2 \\ &\left(\sum_{i=1}^{i=n} x_i\right)^2 = (x_1 + x_2 + \cdots + x_n)^2 \end{aligned}\right\} \tag{2.15}$$

Shorthand versions of the summation notation leave out the upper limit of the summation, and sometimes the lower limit as well. This is done in those situations where *all* of the terms, and not just some subset of them, are to be summed. The following are all equivalent:

$$\sum_{i=1}^{i=n} x_i = \sum_{i}^{n} x_i = \sum_{i} x_i = \sum x_i \tag{2.16}$$

It should also be recognized that the letter "i" is used in this notation simply as an indicator (to indicate which observations or terms to sum); we could just as easily use any other letter:

$$\sum_{i=1}^{i=n} x_i = \sum_{k=1}^{k=n} x_k \tag{2.17}$$

In each case we find the sum by adding up all of the n observations. In fact, we often have use for more than one summation indicator. Double summations are required when we want to denote the sum of all of the observations in a table. A table of commuting flows, such as the one in Table 2.1, indicates the origins and destinations of individuals. The value of any cell is denoted x_{ij} and this refers to the number of commuters from origin i who

Table 2.1 **Hypothetical commuting data**

	Destination		
Origin	**1**	**2**	**3**
1	130	40	50
2	20	100	10
3	30	20	100

commute to destination j. The number of commuters going to destination j from all origins is $\sum_{i=1}^{i=n} x_{ij}$ (where there are n transportation zones), and the number of commuters leaving origin i for all destinations is $\sum_{j=1}^{j=n} x_{ij}$. The total number of commuters is designated by the double summation, $\sum_{i=1}^{n}\sum_{j=1}^{n} x_{ij}$. Using the data in Table 2.1, for example, we find that $\sum_i x_{i2} = 160$, $\sum_j x_{1j} = 220$, and $\sum_i \sum_j x_{ij} = 500$.

Whereas the summation notation refers to the addition of terms, the product notation applies to the multiplication of terms. It is denoted by the capital Greek letter Π, and is used in the same way as the summation notation. For example,

$$\prod_{i=1}^{n}(x_i + y_i) = (x_1 + y_1)(x_2 + y_2)\cdots(x_n + y_n) \tag{2.18}$$

The *factorial* of a positive integer, n, is equal to the product of the first n integers. Surprisingly perhaps, factorials are denoted by an exclamation point. Thus

$$5! = 5 * 4 * 3 * 2 * 1 = 120 \tag{2.19}$$

Note that we could express factorials in terms of the product notation:

$$n! = \prod_{i=1}^{i=n} i \tag{2.20}$$

There is also a convention that $0! = 1$; factorials are not defined for negative integers or for nonintegers.

Factorials arise in the calculation of *combinations*. Combinations refer to the number of possible outcomes that particular probability experiments may have (see Section 2.2).

Specifically, the number of ways that r items may be chosen from a group of n items is denoted by $\binom{n}{r}$, and is equal to

$$\binom{n}{r} = \frac{n!}{r!(n-r)!} \tag{2.21}$$

For example,

$$\binom{5}{2} = \frac{5!}{2!3!} = \frac{120}{2 * 6} = 10 \tag{2.22}$$

What does this mean? If, for example, we group income into $n=5$ categories, then there are ten ways to choose two of them. If we label the five categories (a) through (e), then the ten possible combinations of two income categories are *ab*, *ac*, *ad*, *ae*, *bc*, *bd*, *be*, *cd*, *ce*, and *de*.

2.1.3 Examples

$$6! = 6 * 5 * 4 * 3 * 2 * 1 = 720 \tag{2.23}$$

$$\prod_{i=1}^{i=4} i^2 = 1^2 2^2 3^2 4^2 = 576 \tag{2.24}$$

$$34 + (26/13) * 12 = 58 \tag{2.25}$$

Now let $a=3$, and let the values of a set ($n=3$) of x and y values be $x_1=4$, $x_2=5$, $x_3=6$, $y_1=7$, $y_2=8$, and $y_3=9$. Then

$$\left.\begin{aligned}
&\sum ax_i = 3(4+5+6) = 45\\
&\sum_{i=1}^{2} x_i y_i = (4)(7) + (5)(8) = 68\\
&\sum x_i^3 = 4^3 + 5^3 + 6^3 = 405\\
&\bar{x} = \frac{\sum x_i}{n} = \frac{4+5+6}{3} = 5\\
&s_y^2 = \frac{\sum (y_i - \bar{y})^2}{n-1} = \frac{(7-8)^2 + (8-8)^2 + (9-8)^2}{3-1} = 1
\end{aligned}\right\} \tag{2.26}$$

Note that the sum of products does not necessarily equal the product of sums:

$$\begin{aligned}
\sum x_i y_i &= (4 * 7) + (5 * 8) + (6 * 9) = 122\\
&\neq \sum x_i \sum y_i = (4+5+6) * (7+8+9) = 360
\end{aligned} \tag{2.27}$$

2.2 Sample Spaces, Random Variables, and Probabilities

Suppose we are interested in the likelihood that current residents of a suburban street are new to the neighborhood during the past year. To keep the example manageable, we shall assume that just four households are asked about their duration of residence. There are several possible questions that may be of interest. We may wish to use the sample to estimate the probability that residents of the street moved to the street during the past year. Or we may want to know whether the likelihood of moving onto that street during the past year is any different than it is for the entire city.

This problem is typical of statistical problems in the sense that it is characterized by the *uncertainty* associated with the possible outcomes of the household survey. We may think of the survey as an experiment of sorts. The experiment has associated with it a *sample space*, which is the set of all possible outcomes. Representing a recent move with a "1" and representing longer-term residents with a "0", the sample space is enumerated in Table 2.2. These sixteen outcomes represent all of the possible results from our survey. The individual outcomes are sometimes referred to as *simple events* or *sample points*.

Random variables are functions defined on a sample space. This is a rather formal way of saying that associated with each possible outcome is a quantity of interest to us. In our example, we are unlikely to be interested in the individual responses, but rather the total number of households that are newcomers to the street. Portrayed in Table 2.3 is the sample space with the variable of interest, the number of new households, given in parentheses.

In this instance, the random variable is said to be *discrete*, since it can take on only a finite number of values (namely, the non-negative integers 0–4). Other random variables are *continuous* – they can take on an infinite number of values. Elevation, for example, is a continuous variable.

Associated with each possible outcome in a sample space is a *probability*. Each of the probabilities is greater than or equal to zero, and less than or equal to one. Probabilities may be thought of as a measure of the likelihood or relative frequency of each possible outcome. The sum of the probabilities over the sample space is equal to one.

There are numerous ways to assign probabilities to the elements of sample spaces. One way is to assign them on the basis of relative frequencies. Given a description of the current weather pattern, a meteorologist may note that in 65 out of the last 100 times that such a pattern prevailed there was measurable

Table 2.2 **The sixteen possible outcomes on a sample of four residents**

0000	0100	1000	1100
0001	0101	1001	1101
0010	0110	1010	1110
0011	0111	1011	1111

Table 2.3 **Possible outcomes, with the number of new households in parentheses**

0000 (0)	0100 (1)	1000 (1)	1100 (2)
0001 (1)	0101 (2)	1001 (2)	1101 (3)
0010 (1)	0110 (2)	1010 (2)	1110 (3)
0011 (2)	0111 (3)	1011 (3)	1111 (4)

precipitation the next day. The possible outcomes – rain or no rain tomorrow – are assigned probabilities of 0.65 and 0.35, respectively, on the basis of their relative frequencies.

Another way to assign probabilities is on the basis of subjective beliefs. The description of current weather patterns is a simplification of reality, and may be based upon only a small number of variables such as temperature, wind speed and direction, barometric pressure, etc. The forecaster may, partly on the basis of other experience, assess the likelihoods of precipitation and no precipitation as 0.6 and 0.4, respectively.

Yet another possibility for the assignment of probabilities is to assign each of the n possible outcomes a probability of $1/n$. This approach assumes that each sample point is equally likely, and it is an appropriate way to assign probabilities to the outcomes in special kinds of experiments. If, for example, we flipped four coins, and let "0" represent "heads" and "1" represent "tails," there would be sixteen possible outcomes (identical to the sixteen outcomes associated with our survey of the four residents above). If the probability of heads is 1/2, and if the outcomes of the four tosses are assumed independent from one another, the probability of any particular sequence of four tosses is given by the product $1/2 \times 1/2 \times 1/2 \times 1/2 = 1/16$. Similarly, if the probability that an individual resident is new to the neighborhood is 1/2, we would assign a probability of 1/16 to each of the sixteen outcomes in Table 2.2.

Note that if the probability of heads differs from 1/2, the sixteen outcomes will not be equally likely. If the probability of heads or the probability that a resident is a newcomer is denoted by p, the probability of tails and the probability the resident is *not* a newcomer is equal to $(1 - p)$. In this case, the probability of a particular sequence is again given by the product of the likelihoods of the individual tosses. Thus the likelihood of "1001" (or "HTTH" using H for heads and T for tails) is equal to $p \times (1 - p) \times (1 - p) \times p = p^2(1 - p)^2$.

2.3 The Binomial Distribution

Returning to the example of whether the four surveyed households are newcomers, we are more interested in the random variable defined as the number of new households than in particular sample points. If we want to know the likelihood of receiving two "successes," or two new households out of a survey of four, we must add up all of the probabilities associated with the relevant sample points. In Table 2.4 we use an "*" to designate those outcomes where two households among the four surveyed are new ones.

If the probability that a surveyed household is a new one is equal to p, the likelihood of any particular event with an "*" is $p^2(1 - p)^2$. Since there are six such possibilities, the desired probability is $6p^2(1 - p)^2$.

Table 2.4 **Asterisked outcomes, indicating outcomes of interest**

0000	0100	1000	1100*
0001	0101*	1001*	1101
0010	0110*	1010*	1110
0011*	0111	1011	1111

Note that we have assumed that the probability p is constant across households, and also that households behave independently. These assumptions may or may not be realistic. Different types of household might have different values of p – for example, those who live in bigger houses may be more (or less) likely to be newcomers. The responses received from nearby houses may also not be independent. If one respondent was a newcomer, it might make it more likely that a nearby respondent is also a newcomer (if for example, a new row of houses has just been constructed).

Under these assumptions, the number of households who are newcomers is a *binomial variable*, and the probability that it takes on a particular value is given by the *binomial distribution*. We can find the probability that the random variable, designated X, is equal to 2, using the binomial formula

$$p(X=2) = \binom{4}{2}p^2(1-p)^2 = 6p^2(1-p)^2 \tag{2.28}$$

The binomial coefficient provides a means of counting the number of relevant outcomes in the sample space:

$$\binom{4}{2} = \frac{4!}{2!2!} = \frac{24}{(2)(2)} = 6 \tag{2.29}$$

The binomial distribution is used whenever (a) the process of interest consists of a number (n) of independent trials (in our example, the independent trials were the independent responses of the $n=4$ residents, (b) each trial results in one of two possible outcomes (e.g., a newcomer, or not a newcomer), and (c) the probability of each outcome is known, and is the same for each trial; these probabilities are designated p and $1-p$. Often the outcomes of trials are labelled "success" with probability p and "failure" with probability $1-p$. Then the probability of x successes is given by the binomial distribution

$$p(X=x) = \binom{n}{x}p^x(1-p)^{n-x} \tag{2.30}$$

You should recognize that, for given values of n and p, we can generate a histogram by using this formula to generate the expected frequencies associated with different values of x. The histogram is also known as the binomial probability distribution, and it reveals how likely particular outcomes are. For example, suppose that the probability that a surveyed resident is a newcomer

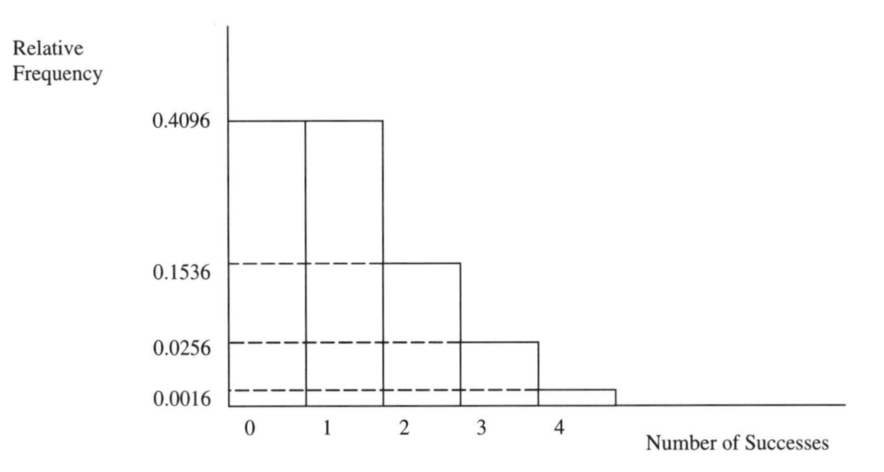

Figure 2.1 **Binomial distribution with $n = 4$, $p = 0.2$**

to the neighborhood is $p = 0.2$. Then the probability that our survey of four residents will result in a given number of newcomers is

$$\left.\begin{aligned} p(X=0) &= \tbinom{4}{0}.2^0.8^4 = .4096 \\ p(X=1) &= \tbinom{4}{1}.2^1.8^3 = .4096 \\ p(X=2) &= \tbinom{4}{2}.2^2.8^2 = .1536 \\ p(X=3) &= \tbinom{4}{3}.2^3.8^1 = .0256 \\ p(X=4) &= \tbinom{4}{4}.2^4.8^0 = .0016 \end{aligned}\right\} \tag{2.31}$$

The probabilities may be thought of as relative frequencies. If we took repeated surveys of four residents, 40.96% of the surveys would yield no newcomers, 40.96% would reveal one newcomer, 15.36% would reveal two newcomers, 2.56% would yield three newcomers, and 0.16% would result in four newcomers. Note that the probabilities or relative frequencies sum to one. The binomial distribution depicted in Figure 2.1 portrays these results graphically. If we multiplied the vertical scale by n, the histogram would represent absolute frequencies expected in each category.

2.4 The Normal Distribution

The most common probability distribution is the *normal distribution*. Its familiar symmetric, bell-shaped appearance is shown in Figure 2.2. The normal distribution is a continuous one – instead of a histogram with a finite number of vertical bars, the relative frequency distribution is continuous. You can think of it as a histogram with a very large number of very narrow vertical bars. The vertical axis is related to the likelihood of obtaining particular x values. As with all frequency distributions, the area under the curve between

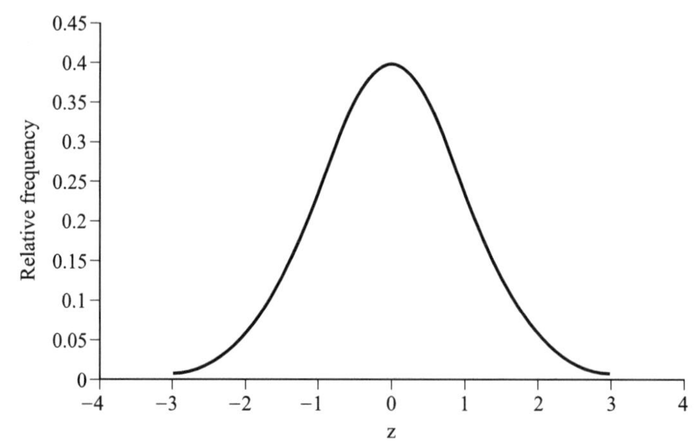

Figure 2.2 **The normal distribution**

any two x values corresponds to the probability of obtaining an x value in that range. The total area under the curve is equal to one.

The normal distribution arises in a variety of contexts and is related to a variety of underlying processes. One way in which the normal distribution arises is through an approximation to binomial processes. Suppose that instead of interviewing four residents, we interviewed 40. We could still use the binomial distribution to evaluate the probability that eleven or fewer households were new to the neighborhood, but that would entail a long, tedious calculation involving large factorials:

$$\begin{aligned} p(X \leq 11) &= p(X = 0) + p(X = 1) + \cdots + p(X = 11) \\ &= \binom{40}{0}.2^0.8^{40} + \cdots + \binom{40}{11}.2^{11}.8^{29} \end{aligned} \tag{2.32}$$

When the sample size is large, the binomial distribution is approximately the same as a normal distribution which has a mean of np and a variance of $np(1-p)$. In our example, we would expect a mean of $np = (40)(0.2) = 8$ residents to indicate that they were newcomers. The variance, $np(1-p) = 40(0.2)(0.8) = 6.4$, represents the variability we would expect in a summary of the results produced by many people who went out and surveyed 40 households.

The probability that eleven or fewer residents are newcomers, $p(X \leq 11)$, may be determined by the shaded area under the normal curve shown in Figure 2.3. The areas under normal curves are given in tables such as that found in Table A.2 in Appendix A. Since variables with normal distributions may have an infinite number of possible means and standard deviations, normal tables are standardized, and they display the areas under normal distributions that have a mean of zero and a standard deviation of one. Before using a normal table, we must transform our data so that it has a mean of zero and a standard deviation of one. This is achieved by converting the data into z-scores.

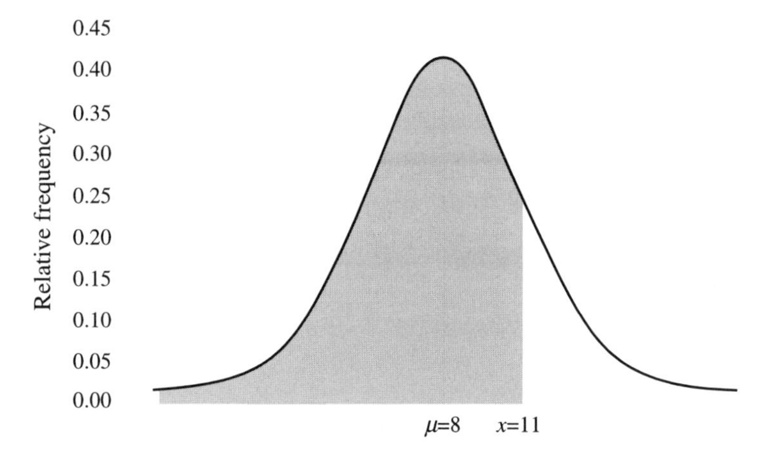

Figure 2.3 **Probability of $X < 11$**

For our example, we convert $x = 11$ into a z-score by first subtracting the mean and then dividing the result by the standard deviation:

$$z = \frac{11 - 8}{\sqrt{6.4}} = 1.19 \tag{2.33}$$

We now find the probability that $z < 1.19$ from the normal table; it is equal to 0.8830. To be a bit more precise, we would account for the fact that our variable of interest is a discrete one. The vertical bar associated with $x = 11$ on a histogram of the binomial distribution would stretch from $x = 10.5$ to $x = 11.5$. We can get a better approximation by finding the probability that $x < 11.5$. Converting $x = 11.5$ to a z-score yields $z = 1.38$, and from the normal table the probability that $z < 1.38$ is 0.9162. For comparison, the binomial formula leads to a probability of 0.9125.

Whether the binomial distribution may be approximated well by the normal distribution depends upon the values of n and p. The binomial distribution is only truly symmetric when $p = 0.5$, and so, if p is near either 0 or 1, the normal approximation may not be accurate. The normal approximation also improves as n becomes large. A common rule-of-thumb is that np should be greater than 5. In our example np was equal to 8, and the approximation was fairly accurate.

As we will see in the next subsection, the normal distribution also characterizes the distribution of sample means. In particular, sample means have a normal distribution, with mean equal to the true population mean (μ) and variance equal to σ^2/n, where σ^2 is the population variance and n is the sample size. If we repeatedly took samples of size n from a population, and then made a histogram of all of our sample means, the histogram would have the appearance of a normal distribution with mean μ and variance equal to σ^2/n.

2.5 Confidence Intervals for the Mean

The *central limit theorem* tells us something about the nature of sample means. Any time we sum a large number of independent, identically distributed variables, the central limit theorem tells us that the sum will have a normal, bell-shaped curve for its frequency distribution.

Using the data on commuting times from Chapter 1, we begin by summing the commuting times (and then simply dividing by a constant, n, to obtain the mean). "Independent" implies that one individual's commuting time is unrelated to the commuting time of other individuals. "Identically distributed" means that each individual commuting time comes from the same frequency distribution. In other words, there are not separate frequency distributions that govern separate subcategories of the population. Under these conditions, we would find that the frequency distribution of sample means (which could be constructed if we had the results of many surveys, each with its own sample mean) would follow a normal distribution.

Furthermore, the normal, bell-shaped curve representing the frequency distribution of the means will have a mean equal to the true mean, μ. Although it is unlikely that our own sample mean will be equal to the true mean, it is certainly reassuring to know that, with a large number of people repeating our survey of thirty individuals, the average of all of the sample means would be equal to the true mean.

Finally, we know something about the variability that we will observe among the sample means collected. In particular, the sample means will have a variance equal to σ^2/n . This is consistent with one's intuition that sample means will display more variability when the original data are inherently variable; high values of σ^2 will lead to high values of σ^2/n. If we were all to take large surveys (i.e., make n larger), the resulting distribution of sample means would also display less variability.

Summarizing, we know that if others repeated our survey, and if we made a histogram using all of the many sample means that were collected, the histogram would have a roughly normal, bell-shaped appearance, the mean of the sample means would provide an estimate of the true mean, and the variance of the sample means would be equal to σ^2/n.

Since we know something about the distribution of sample means, we can make statements about how confident we are that the true mean is within a given interval about our sample mean. For a normal distribution, 95% of the observations lie within about two standard deviations (actually 1.96 standard deviations) of the mean. This can be verified using the standard normal table (Table A.2 in Appendix A), which reveals that the probability of a z-score with absolute value less than 1.96 is 0.95. This implies that our individual sample mean should, 95% of the time, lie within $\pm 1.96s/\sqrt{n}$ of the true mean, μ:

$$\mathrm{pr}\left[\left(\mu - 1.96\frac{s}{\sqrt{n}}\right) \leq \bar{x} \leq \left(\mu + 1.96\frac{s}{\sqrt{n}}\right)\right] = 0.95 \tag{2.34}$$

The probability that $\bar{x}$ lies in the range described in parentheses is 0.95. Rearranging allows us to construct a confidence interval around our sample mean:

$$\mathrm{pr}\left[\left(\bar{x} - 1.96\frac{s}{\sqrt{n}}\right) \leq \mu \leq \left(\bar{x} + 1.96\frac{s}{\sqrt{n}}\right)\right] = 0.95 \qquad (2.35)$$

This tells us that 95% of the time the true mean should lie within $\pm 1.96s/\sqrt{n}$ of the sample mean. A 90% confidence interval could be constructed by recognizing that the true mean would lie within 1.645 standard deviations of the mean 90% of the time. The value of 1.645 comes from the standard normal z-table (Table A.2 in Appendix A); it is the value of z associated with 5% of the area under each of the two tails of the distribution. It is also fairly common to use 99% confidence intervals, constructed by adding and subtracting $2.58s/\sqrt{n}$ to the sample mean.

More generally, a $(1 - \alpha)\%$ confidence interval around the sample mean is

$$\mathrm{pr}\left[\left(\bar{x} - z_\alpha\frac{s}{\sqrt{n}}\right) \leq \mu \leq \left(\bar{x} + z_\alpha\frac{s}{\sqrt{n}}\right)\right] = 1 - \alpha \qquad (2.36)$$

where z_α is the value taken from the z-table that is associated with a fraction α of the weight in the tails (and therefore $\alpha/2$ is the area in *each* tail).

Before proceeding to apply this to our commuting data example, we must consider one more factor. The central limit theorem applies when the sample size is "large;" only then will the distribution of means possess a normal distribution. When the sample size is not "large," the frequency distribution of the sample means has what is known as the t-distribution; it is symmetric, like the normal distribution, but it has a slightly different shape. The areas under the t-distribution are given in Table A.3 in Appendix A. With a sample size of $n = 30$, 95% confidence intervals are constructed using $t = 2.045$, instead of the value of $z = 1.96$ used above for the normal distribution.

For the commuting data, the 95% confidence interval around the mean is therefore

$$\mathrm{pr}\left[\left\{21.93 - \frac{2.045(14.43)}{\sqrt{30}}\right\} \leq \mu \leq \left\{21.93 + \frac{2.045(14.43)}{\sqrt{30}}\right\}\right] = 0.95 \quad (2.37)$$

and thus we are 95% sure that the true mean is within plus or minus $2.045\times(14.43)/\sqrt{30} = 5.39$ of our sample mean of 21.93. The 95% confidence interval for the mean may be stated as (16.54, 27.32). More precisely, 95% of confidence intervals constructed from samples in this way will contain the true mean.

2.6 Probability Models

Probability is used as a basis for statistical inference. In college mathematics course sequences probability comes before statistics, since probability concepts form the foundation of statistical tests and statistical inference. In addition to forming the basis of standard statistical tests, probability is used to develop models of geographic processes. Though the primary emphasis of this book is on the use of probability in statistical inference, in this section we provide some examples of probability modeling.

The outline of the scientific method in the previous chapter indicated that models are used as simplifications of reality. A simplified view of reality permits one to focus on the nature of the relationships between key variables. Uncertainty and probability are concepts that are central to the construction of many models in geography. A model that is particularly useful in illustrating both the nature of models and the manner in which probability is central is the *intervening opportunities model* (Stouffer 1940).

2.6.1 The Intervening Opportunities Model

The intervening opportunities model was originally used in the context of migration, but has since been used more widely in the field of transportation. The conceptual foundation rests on the idea that the movement behavior of individuals in space obeys the principle of least effort – individuals will consider opportunities that are closest to them first, and if they find them unacceptable they will go on to the next closest opportunity or opportunities.

This conceptual foundation is quite easy to develop into a probability model that indicates how individual travel behavior might be organized. Suppose that an individual consumer is considering a purchase, and that there are several alternative stores in the vicinity where a purchase might be made. If there are n stores, we can arrange them in order of their distances from the individual. Let us call the store closest to the individual "store 1," and the one that is furthest away "store n." The intervening opportunities model makes just two assumptions – that (a) individuals consider opportunities sequentially, in order of distance; and (b) individuals consider each opportunity, and find each opportunity acceptable with constant probability L.

These two assumptions imply that our individual starts by considering the closest opportunity. The probability that it is acceptable is L, so we may write the probability of stopping at the closest store as

$$p(X = 1) = L \tag{2.38}$$

We use X to denote the random variable we are interested in – namely, the number of the store the individual purchases from. The probability that the individual finds the closest opportunity *unacceptable* is $1 - L$; if this occurs the person goes on to the second closest opportunity, and accepts it with probability L. The probability the individual ends up purchasing at store 2 is therefore the product of these two independent terms, representing the likelihood of rejecting the first opportunity and accepting the second:

$$p(X = 2) = (1 - L)L \tag{2.39}$$

Similar reasoning implies that the probability of rejecting the first two opportunities and accepting the third is

$$p(X = 3) = (1 - L)(1 - L)L = (1 - L)^2 L \tag{2.40}$$

In general, the probability of accepting opportunity j is equal to the opportunity of rejecting the first $j - 1$ opportunities, multiplied by the likelihood of accepting the jth opportunity:

$$p(X = j) = (1 - L)^{j-1} L \tag{2.41}$$

In this model, the variable X is a *geometric* variable that is characterized by a downward-sloping histogram. For example, if $L = 0.5$, the probabilities are $p(X = 1) = 0.5$, $p(X = 2) = 0.25$, $p(X = 3) = 0.125$, $p(X = 4) = 0.0625$, etc. Note how this simple model captures one of the most important of all geographical concepts – namely, that geographic interaction declines with increasing distance. Although the model is an oversimplification in many respects (e.g., the probability of accepting any given opportunity is probably *not* constant in most contexts), it does capture the most important feature of spatial interaction.

Once we have set out the main features of a probability model such as the intervening opportunities model, other questions naturally arise. Some of the questions that might arise here are:

(1) Do the probabilities add to one, as they should?
(2) Where does L come from?
(3) What if we do not have data on the exact distances of the opportunities, and the data are arranged into zones around the origin?

We now consider each of these questions in turn.

Do the probabilities add to one? A geometric random variable is used whenever the variable of interest is the number of the trial on which the first success occurs. Here the "trials" are the opportunities, and "success" refers to the selection of a particular opportunity. If the variable is allowed to take for its value any positive integer $(1, 2, \ldots)$, then the probabilities will add to one. In the intervening opportunities model, an individual will not have an infinite number of opportunities to consider, and therefore the probabilities will sum to less than one. That is,

$$\sum_{i=1}^{i=n<\infty} p(X = i) < 1 \tag{2.42}$$

Consequently, to be more precise, we should adjust our probabilities accordingly. We may do so by dividing by their sum, to ensure that they add to one:

$$p(X = j) = \frac{(1 - L)^{j-1} L}{\sum_{i=1}^{n} (1 - L)^{i-1} L} \tag{2.43}$$

If either (a) n is large, (b) L is large, or (c) n is large *and* L is large, then such an adjustment is unnecessary, since the denominator in the previous equation will be close to one.

Where does L come from? Suppose we have a set of data indicating the proportion of people leaving a particular residential origin who end up at each of the n destinations. We would like to choose (i.e., estimate) L in a way that is consistent with what we observe. That is, since we have the freedom to estimate L, clearly we should do so in a way that mimics as closely as possible our observed data.

There are many alternative approaches to choosing L, and here we will illustrate several of them. Suppose we observe the following proportions of people leaving an origin for one of the six potential stores in the area (arranged in terms of increasing distance away from the origin): $p_{\text{obs}}(X=1)=0.55$, $p_{\text{obs}}(X=2)=0.3$, $p_{\text{obs}}(X=3)=0.1$, $p_{\text{obs}}(X=4)=0.05$, $p_{\text{obs}}(X=5)=0.0$, and $p_{\text{obs}}(X=6)=0.0$. Most individuals go to the closest store, and no one goes from our origin to the two stores that are farthest away. The nature of the observations therefore suggests that the intervening opportunities model might work well in replicating observed travel behavior.

One way to choose L would be to simply try several values and see which one works best. We could define "best" in different ways, but let us use the sum of the squared deviations between observed and predicted values. Thus if we try $L=0.5$, our criterion, the sum of squared deviations is

$$\sum_{i=1}^{i=n} (p_{\text{obs}}(i) - \hat{p}(i))^2 = (.55 - .5)^2 + (.3 - .25)^2 + (.1 - .125)^2 + (.05 - .0625)^2 + (0 - .03125)^2 + (0 - .01563)^2 = .0145 \tag{2.44}$$

where the "^" indicates the probability predicted by the model (where we have used the simple form of the model given by Equation 2.41, which may be justified not by the fact that n is large but rather by the fact that L is likely to be large, since observed interaction falls off so sharply with distance). If we repeat this for many values of L, we obtain the graph in Figure 2.4, which shows that the deviations are minimized (and hence the fit of the model is best) at a value of $\hat{L}=0.561$. Again, the "^" notation is used to indicate that L is an estimate of the true, unknown value of L. With $\hat{L}=0.561$, the predicted probabilities are

$$\hat{p}(X=1)=0.561, \quad \hat{p}(X=2)=0.246, \quad \hat{p}(X=3)=0.108,$$
$$\hat{p}(X=4)=0.047, \quad \hat{p}(X=5)=0.021, \quad \hat{p}(X=6)=0.009 \qquad (2.45)$$

and these are quite close to the observed values.

An alternative way to estimate L is to make use of the fact that the mean of a geometric random variable is equal to the reciprocal of the probability of success (or, here, the reciprocal of L, the probability of accepting a given opportunity). The mean destination of our sampled population is 1.7. To see this, suppose that 100 people left the origin. Fifty-five would stop at destination 1, 30 at destination 2, 10 at destination 3, and 5 at destination 4. If we make a list of the destination numbers for each of these 100 individuals, the total will be $(55+(30)(2)+(10)(3)+(4)(5)=165)$, and dividing by the total number of people in the sample yields $165/100=1.65$. Thus we have $\hat{L}=1/1.65=0.606$, which is similar to our previous estimate of L.

What if the opportunity data are organized into zones? A common way of using the intervening opportunities model is to organize the potential destinations into zones around the origin. Transportation planners do not often need to know the number of individuals arriving at each destination; they are satisfied with more aggregate estimates of the number arriving within defined transportation zones. We may specify the number of opportunities in destination zone j as d_j, and now arrange the zones in order of increasing

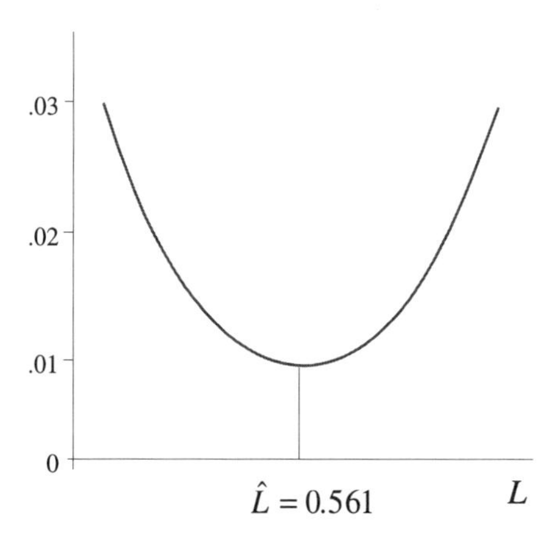

Figure 2.4 **Sum of squared errors as a function of *L***

distance around the origin. The probability of stopping somewhere in the zone closest to the origin is equal to the probability of accepting an opportunity somewhere (anywhere) within it. In turn, this may be thought of as one minus the probability of *not* finding any of zone 1's opportunities acceptable:

$$p(X = 1) = 1 - (1 - L)^{d_1} \tag{2.46}$$

where $p(X = 1)$ now refers to the probability of stopping in zone 1. The probability of stopping in zone 2 is equal to the probability of going beyond zone 1, minus the probability of going beyond zone 2:

$$p(X = 2) = (1 - L)^{d_1} - (1 - L)^{d_1 + d_2} \tag{2.47}$$

In general, the probability of stopping in zone j is equal to the probability of going beyond zone $j - 1$, minus the probability of going beyond zone j:

$$p(X = j) = (1 - L)^{\sum_{i=1}^{j-1} d_i} - (1 - L)^{\sum_{i=1}^{j} d_i} \tag{2.48}$$

L may be determined by minimizing the sum of squared errors, as described in the previous subsection. For example, suppose that $d_1 = 5$, $d_2 = 4$, $d_3 = 4$, $p_{obs}(1) = .5$, $p_{obs}(2) = .4$, and $p_{obs}(3) = .1$. By trying alternative L values we find that $L = 0.132$ minimizes the sum of squared errors

$$\sum_{i=1}^{3} [p_{obs}(i) - \hat{p}(i)]^2 \tag{2.49}$$

where the predicted values $\hat{p}(i)$ are found from Equations 2.46 to 2.48.

2.6.2 *A Model of Migration*

It is possible to construct a simple model of the movement of people for a regional system that has been divided into n subareas. For illustrative purposes, we will look at the $n = 2$ case of movement between central city and suburbs. We focus only on the redistribution of people who are alive and living in the study region throughout the study period.

Suppose that each year 20% of central city residents move to the suburbs and 15% of all suburban residents move to the central city. If we start with 10 000 residents in each location, after the first year we have, for the suburban and central city populations

$$\left.\begin{aligned} P_{sub} &= .85(10\,000) + .2(10\,000) = 10\,500 \\ P_{cc} &= .8(10\,000) + .15(10\,000) = 9500 \end{aligned}\right\} \tag{2.50}$$

Assuming that these probabilities of movement remain constant over time, the population at the end of year t can be written in terms of the populations at the end of the previous year:

$$\left.\begin{aligned} P_{\text{sub}}(t) &= .85P_{\text{sub}}(t-1) + .2P_{\text{cc}}(t-1) \\ P_{\text{cc}}(t) &= .8P_{\text{cc}}(t-1) + .15P_{\text{sub}}(t-1) \end{aligned}\right\} \tag{2.51}$$

For example, the populations at the end of the second year are

$$\left.\begin{aligned} P_{\text{sub}}(t) &= .85(10\,500) + .2(9500) = 10\,825 \\ P_{\text{cc}}(t) &= .8(9500) + .15(10\,500) = 9175 \end{aligned}\right\} \tag{2.52}$$

Note that the total population remains fixed at 20 000. This simple model is known as the *Markov model*, and it provides a useful short-run method for projecting the migration component of population change.

The model also has interesting long-run properties. If the model is allowed to run for a long time, the regional populations approach a constant, equilibrium value, i.e.,

$$\left.\begin{aligned} P_{\text{sub}}(t) &= .85P_{\text{sub}}(t) + .2P_{\text{cc}}(t) \\ P_{\text{cc}}(t) &= .8P_{\text{cc}}(t) + .15P_{\text{sub}}(t) \end{aligned}\right\} \tag{2.53}$$

The equilibrium populations may be determined by taking either equation from 2.53 together with the fact that the total population is fixed at 20 000. For example,

$$\left.\begin{aligned} P_{\text{sub}}(t) &= .85P_{\text{sub}}(t) + .2P_{\text{cc}}(t) \\ P_{\text{sub}}(t) &+ P_{\text{cc}}(t) = 20\,000 \end{aligned}\right\} \tag{2.54}$$

is a set of two equations and two unknowns. It may easily be solved to yield $P_{\text{sub}} = 11\,429$ and $P_{\text{cc}} = 8571$. Thus, if current probabilities of movement do not change, 4/7 of the regional population will reside in the suburbs and 3/7 will reside in the central city. These equilibrium populations depend only upon the probabilities of movement between subareas; they do not depend upon the initial, subarea populations. Although in reality probabilities of movement do not remain constant for such long periods of time, the long-run equilibrium provides a useful (moving) target toward which the population distribution is heading.

The Markov model provides a good illustration of how elementary probability concepts may be used to model an important process. In this case, the model provides both useful short-range forecasts and understandable long-range consequences. For more details on this model, see Rogers (1975).

2.6.3 The Future of the Human Population

How long will the human species survive? This basic question has been the subject of much debate. Attention has been given to factors such as the rate at which we are deplenishing nonrenewable resources, as well as how many people the earth can support (Cohen 1995).

An interesting approach, using a simple probability argument, to estimating the survival of the human species has been suggested by Gott (1993). Imagine a list of all humans who have ever lived, or will ever live. Since there is no reason to suppose that we occupy a special place on this list, there is only a 5% chance that we are listed among either the first 2.5% on the list or the last 2.5% on the list. As Gott notes:

> Assume that you are located randomly on the chronological list of human beings. If the total number of intelligent individuals in the species is a positive integer $N_{tot} = N_{past} + 1 + N_{future}$, where N_{past} is the number of intelligent individuals born before a particular intelligent observer and N_{future} is the number born after, then we expect N_{past} to be the integer part of the number rN_{tot}, where r is a random number between 0 and 1.

We know that N_{past} is approximately 70 billion. If N_{future} turns out to be greater than about 2.8 trillion, that would mean we would be on the first 2.5% of the chronological list (since 70 billion/2.8 trillion = 0.025). Similarly, if N_{future} turns out to be less than 1.75 billion, that would mean we would be on the last 2.5% of the chronological list (since 1.75 billion/70 billion = 0.025). If we do not occupy a special place on the chronological list, this means that the future population yet to be born is, with 95% confidence, in the range

$$1.75\,\text{billion} \leq N_{future} \leq 2.8\,\text{trillion} \tag{2.55}$$

Using birth rates that are now slightly out-of-date, Gott translates this into a 95% confidence interval for the length of survival of the human species:

$$12\,\text{years} \leq t_{future} \leq 7.8\,\text{million years} \tag{2.56}$$

where t_{future} is the number of additional years that the human species will survive. Readers will probably be comfortable with the upper limit, but will be surprised by the lower limit! Even the upper limit is not large when considering, for example, the length of time that life has been present on the planet.

Different results are obtained when one uses different assumptions. Our species has been in existence for approximately $t_{past} = 200\,000$ years. If we do not occupy a special place on the timeline of the past and future history of the human species, there is a 2.5% chance that we will live in years that are within the first 2.5% of the ultimate timeline, and a 2.5% chance that we will live in years that are within the last 2.5%. If t_{future} is less than 5000 years, this would mean that we would be in the last 2.5% of the timeline (since 5000/200 000 = 0.025), and if t_{future} is greater than 8 million years, this would mean that we would be in the first 2.5% of the timeline (since 200 000/8 million = 2.5%). Hence

$$5000\,\text{years} \leq t_{\text{future}} \leq 8\,\text{million years} \tag{2.57}$$

So we get a little more time at the lower limit with this scenario.

Exercises

1. The probability of a dry summer is equal to 0.3, the probability of a wet summer is equal to 0.2, and the probability of a summer with normal precipitation is equal to 0.5. A climatologist observed the precipitation during three consecutive summers.

(a) Enumerate the sample space, and assign probabilities to each simple event.
(b) What is the probability of observing two dry summers?
(c) What is the probability of observing at least two dry summers?
(d) What is the probability of not observing a wet summer?

2. Snowfall for a location is found to be normally distributed with mean 96 in. and standard deviation 32 in.

(a) What is the probability that a given year will have more than 120 inches of snow?
(b) What is the probability that the snowfall will be between 90 in. and 100 in.?
(c) What level of snowfall will be exceeded only 10% of the time?

3. Let $a = 5$, $x_1 = 6$, $x_2 = 7$, $x_3 = 8$, $x_4 = 10$, $x_5 = 11$, $y_1 = 3$, $y_2 = 5$, $y_3 = 6$, $y_4 = 14$, and $y_5 = 12$. Find the following:

$$\sum x_i$$

$$\sum x_i y_i$$

$$\sum (x_i + a y_i)$$

$$\sum_{i=1}^{3} y_i^2$$

$$\sum_{i=1}^{i=n} a$$

$$\sum_{k} 2(y_k - 3)$$

$$\sum_{i=1}^{i=5} (x_i - \bar{x})(y_i - \bar{y})$$

4. Find $8!/3!$.

5. Find $\binom{10}{5}$.

6. Use the following table of commuting flows to determine the total number of commuters leaving each zone and the total number entering each zone. Also find the total number of commuters. For each answer, also give the correct notation, assuming y_{ij} denotes the number of commuters who leave origin i to go to destination zone j.

	Destination zone			
Origin zone	**1**	**2**	**3**	**4**
1	32	25	14	10
2	14	33	19	9
3	15	27	39	20
4	10	12	20	40

7. The following data represent stream link lengths in a river network (given in meters): 100, 426, 322, 466, 112, 155, 388, 1155, 234, 324, 556, 221, 18, 133, 177, 441.

(a) Find the mean and standard deviation of the link lengths.
(b) Find 90% and 95% confidence intervals for the mean.

8. If the probability that an individual moves outside of his or her county of residence in a given year is 0.15, what is the probability that

(a) less than three out of a sample of ten move outside the county?
(b) at least one moves outside the county?

9. The annual probability that an individual makes an interstate move is 0.03. What is the probability that at least two out of ten people will make an interstate move next year?

10. Assume that the prices paid for housing within a neighborhood have a normal distribution, with mean \$100 000 and standard deviation \$35 000.

(a) What percentage of houses in the neighborhood have prices between \$90 000 and \$130 000?
(b) What price of housing is such that only 12% of all houses in the neighborhood have lower prices?

11. Assume that the probability that an individual changes residence during the year is 0.21. A survey is taken of 5 individuals.

(a) Write out the elements in the sample space.

(b) What is the probability that three out of five individuals move during the year?
(c) What is the probability that at least one of the five individuals moves during the year?

12. Residents in a community have a choice of six different grocery stores. The proportions of residents observed to patronize each are $p(1) = .4$, $p(2) = .25$, $p(3) = .15$, $p(4) = .1$, $p(5) = .05$, and $p(6) = 0.05$, where the stores are arranged in terms of increasing distance from the residential community. Fit an intervening opportunities model to these data by estimating the parameter L.

13. The annual probability that suburban residents move to the central city is 0.08, while the annual probability that central city residents move to the suburbs is 0.11. Starting with respective populations of 30 000 and 20 000 in the central city and suburbs, forecast the population redistribution that will occur over the next three years. Use the Markov model assumption that the probabilities of movement will remain constant. Also find the long-run, equilibrium populations.

14. The probability that an individual commutes to work by car is 0.9. What is the probability that ten neighbors *all* commute by car? What is the probability that exactly eight of the ten commute by car?

3 Hypothesis Testing and Sampling

LEARNING OBJECTIVES

- Formation and testing of hypotheses
- Hypothesis testing for one- and two-sample tests of means and proportions
- Understanding the distinction between distributions of the variable of interest and distributions of the test statistic
- The special nature of spatial nature, and complications brought about by nonindependence of spatial observations
- Issues associated with sampling

Hypothesis testing is the fundamental way in which inferences about a population are made from a sample. In this chapter, we first focus on the testing of hypotheses involving either one sample or two samples. The latter part of the chapter focuses upon several alternative methods of sampling that may be used to gather information.

3.1 Hypothesis Testing and One-Sample *z*-Tests of the Mean

We will first describe some of the basic concepts of hypothesis testing and statistical inference through an example using a one-sample test involving a mean. Suppose we want to know whether the mean number of weekly shopping trips made by households in a particular neighborhood of an urban area differs from 3.1, which is the corresponding mean for the urban area as a whole. We do not wish to survey all households in the neighborhood to find the desired mean, since that would be too costly (and if we *could* do this, it would be wasteful). Instead, we choose to take a random sample of households. In this example, it is assumed that the value of 3.1, which applies to the entire urban area, is known.

The first step is to set up a *null hypothesis*, where the mean number of shopping trips in the neighborhood is equal to the mean for the entire urban area:

$$H_0 : \mu = 3.1 \tag{3.1}$$

where μ is the hypothesized, true mean for the neighborhood. Null hypotheses are set up in this way, where to accept it will be in keeping with the default option that the neighborhood mean is no different from the hypothesized mean for the entire urban area. This would be a null result. Rejecting the null hypothesis occurs when we find evidence for a significant difference from the hypothesized mean.

The second step is to state an *alternative hypothesis*. Suppose that we are interested in this example because we suspect from other anecdotal evidence that the neighborhood of interest has a high number of shopping trips. In this case, our alternative hypothesis is that the true, unknown mean in the neighborhood is greater than 3.1:

$$H_A : \mu > 3.1 \tag{3.2}$$

This is known as a *one-sided* hypothesis, since we suspect that if the true neighborhood mean *does* differ from 3.1, it will be greater than 3.1 and not less. If on the other hand we had no *a priori* idea about how the neighborhood mean might differ from that for the urban area, we would postulate the following:

$$H_A : \mu \neq 3.1 \tag{3.3}$$

Here we have a *two-sided hypothesis*; if the true neighborhood mean differs from 3.1, it could lie on either side of 3.1. In carrying out statistical tests, we need to recognize that we will be making decisions on the basis of a sample drawn from a larger population. We will never know for certain whether the null hypothesis is true or false. We base our decision on the evidence in favor of or against the hypothesis. If we interview ten households, and the sample mean is 11.7 shopping trips/week, two conclusions are possible. One possibility is that the null hypothesis is true. In this case, the true mean among households in the neighborhood is 3.1, and we have obtained an unusual sample. The other possibility is that the null hypothesis is false. In this event, the true mean among households in the neighborhood is not equal to 3.1. We decide in favor of the null hypothesis if, under H_0, the sample is not *too* unusual; otherwise we will reject H_0. The role of statistics in this case is to inform us regarding precisely how unusual it would be to obtain our sample if the null hypothesis were true.

In the course of this process, it is possible that one of two kinds of error might be made. A *Type I error* refers to rejecting a true hypothesis, whereas a *Type II error* refers to accepting a false hypothesis. The likelihood of making a Type I error is denoted by α and is referred to as the *significance level*. The analyst has control over α and the third step in setting up a statistical test is to choose a significance level. Common values chosen for α are 0.01, 0.05, and 0.10. Though we, of course, wish to keep the likelihood of errors as small as possible, we cannot simply choose α to be exceedingly small. This is because there is an inverse relationship between α and β, the likelihood of making a Type II error. The lower we choose α, the greater the chance that we will accept a false hypothesis. The two columns of Figure 3.1 summarize the four possible outcomes associated with statistical testing. If the null hypothesis is true, we either make a correct decision with probability $1-\alpha$ or an incorrect decision with probability α. If the null hypothesis is false, we either make a correct decision (with probability $1-\beta$), or, with probability β, we make a Type II error.

The fourth step in hypothesis testing is to choose a test statistic and to find the observed value of the test statistic.

For one-sample tests involving the mean, we have sample data $x_1, x_2, \ldots, x_n$. From Chapter 2 we have learned that sample means are normally distributed with mean μ and standard deviation $\sigma/\sqrt{n}$. If we replace the unknown population standard deviation with its sample estimate (s), we can use the test statistic

$$z = \frac{\bar{x}-\mu}{s/\sqrt{n}} \tag{3.4}$$

This test statistic will have, for n greater than about 30, a standard normal distribution with mean 0 and standard deviation 1. The normal distribution is often denoted $N(\mu, \sigma^2)$ which, in this case of the standard normal distribution, is $N(0, 1)$.

Suppose that, for our example, we interview $n = 100$ individuals in the neighborhood and find that the sample mean is 4.2 shopping trips per week, with a sample standard deviation of $s = 5.0$. Then the observed z-statistic is

$$z = \frac{4.2-3.1}{5/\sqrt{100}} = \frac{1.1}{0.5} = 2.2 \tag{3.5}$$

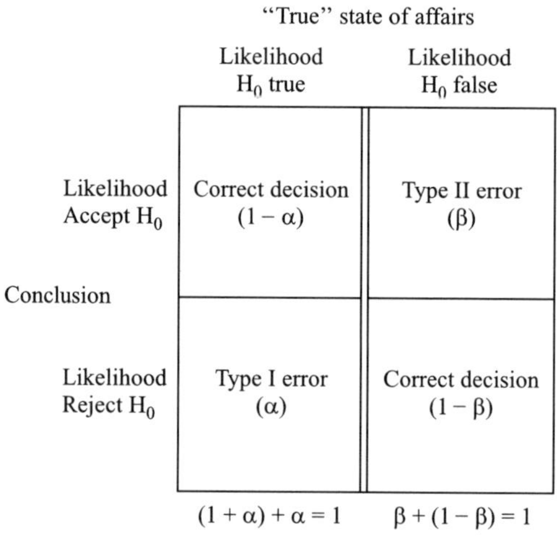

		"True" state of affairs	
		Likelihood H_0 true	Likelihood H_0 false
Conclusion	Likelihood Accept H_0	Correct decision $(1-\alpha)$	Type II error (β)
	Likelihood Reject H_0	Type I error (α)	Correct decision $(1-\beta)$
		$(1+\alpha)+\alpha=1$	$\beta+(1-\beta)=1$

Figure 3.1 **Four outcomes associated with statistical testing**

Intuitively, the z-score is large in absolute value when the sample mean is far from the hypothesized mean, and it is in such cases that we will reject the null hypothesis for two-sided alternative hypotheses.

How large must the test statistic be before we reject H_0? The fifth step in hypothesis testing is to use α and our knowledge of the sampling distribution of the test statistic to determine the *critical value* of the test statistic. Critical values are those values of the test statistic where we are on the knife-edge between acceptance and rejection. If the observed test statistic is slightly to one side of the critical value, we accept H_0; if it is slightly to the other side, we reject H_0. Returning to our example, where the sampling distribution is normal, if we have chosen $\alpha = 0.05$ and the two-sided alternative $H_0 : \mu \neq 3.1$, the critical values of z are equal to -1.96 and 1.96 (Figure 3.2); under H_0, we would expect 5% of all experiments to result in $|Z| > 1.96$. Since our observed value of $2.2 > 1.96$, we reject H_0. How often would we observe a value as high (or higher) as our observed value of 2.2 under H_0? A check of the table of normal distribution probabilities reveals that such an event would occur with probability $(2)0.0139 = 0.0278$ (see Figure 3.3). The value of 0.0278 is referred to as the *p-value*; it tells us how likely a result more extreme than the one we observed would be if the null hypothesis were true.

Note that if we had used the one-tailed alternative $H_A : \mu > 3.1$ we would also have rejected H_0, since the critical value of z (denoted z_{crit}) would have been equal to 1.645 (Figure 3.4).

The steps involved in hypothesis testing are summarized below:

Hypothesis testing

(1) State the null hypothesis, H_0.
(2) State the alternative hypothesis, H_A.
(3) Choose α, the probability of making a Type I error (rejecting a true H_0).
(4) Choose a statistical test, and find the observed test statistic.

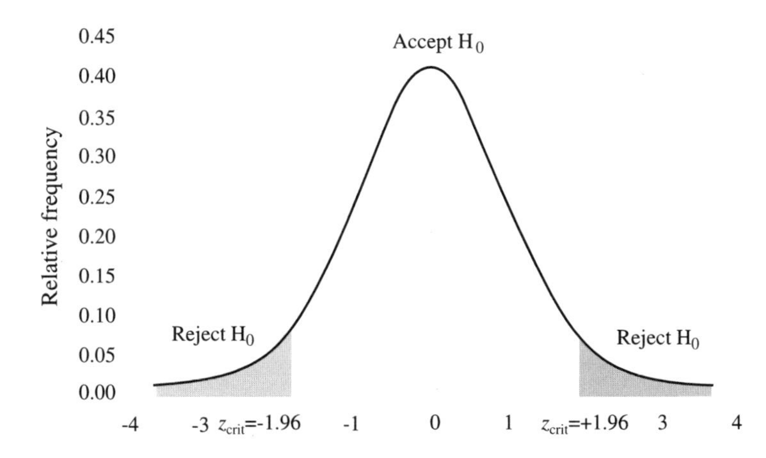

Figure 3.2 Critical regions of the sampling distribution of the difference of means

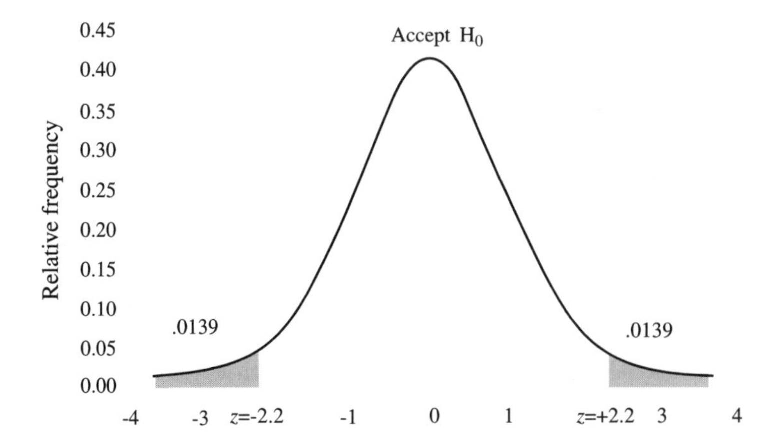

Figure 3.3 **A *p*-value of 0.0278**

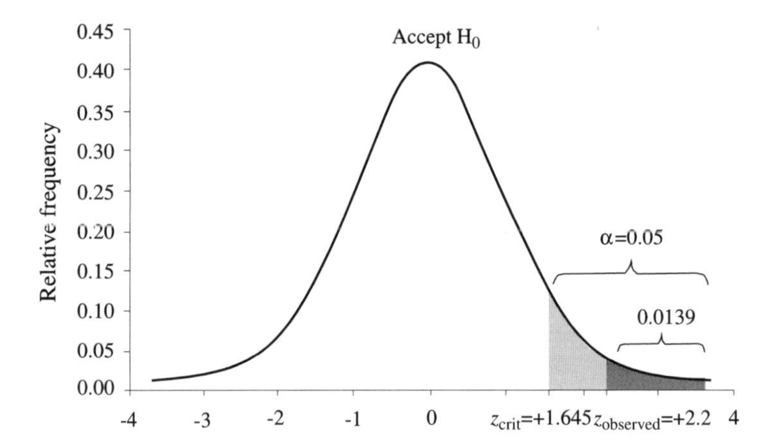

Figure 3.4 **Critical and *p*-values in one-sided test**

(5) Find the critical value of the test statistic to determine which values of the observed statistic will imply rejection of H_0.

(6) Compare the observed test statistic with the critical value of the test statistic, and decide to accept or reject H_0.

In Section 3.4, we will review two-sample tests for differences in means. This will lead naturally into the topic of analysis of variance in Chapter 4, which is concerned with possible differences in means among three or more samples.

3.2 One-Sample *t*-Tests

When the population variance is unknown (as it almost always is; we don't even know the population mean (μ), so how would we know the variance?) and the sample size is small, the sampling distribution of the mean is no longer normal, and hence we should not use the z-statistic. Instead, the sampling distribution of the mean follows a t-distribution with $n-1$ degrees of freedom. Degrees of freedom may be somewhat loosely thought of as the number of observations minus the number of quantities estimated. We have n observations, and we "use up" one degree of freedom to estimate the mean (the sample mean is also used to estimate the sample variance). We have $n-1$ degrees of freedom since, if I gave you the values of $n-1$ observations, knowing the mean, you could calculate the value of the nth observation without being told it.

In addition to the usual assumption that the observations are independent, use of the t-distribution requires the additional assumption that the observations come from a normal distribution.

The t-distribution is similar in shape to the normal distribution, though the tails of the distribution are slightly fatter in comparison with the normal distribution. The t-statistic is found in exactly the same way as the z-statistic:

$$t = \frac{\bar{x} - \mu}{s/\sqrt{n}} \tag{3.6}$$

To test a hypothesis about the mean, we compare our observed t-statistic with the critical value taken from a t-table (see, e.g., Table A.3a in Appendix A) with $n-1$ degrees of freedom.

3.2.1 *Illustration*

Suppose that in our previous example we interviewed $n=20$ people instead of $n=100$, and found $\bar{x}=4.5$ and $s=5.5$. Our test statistic is

$$t = \frac{4.5 - 3.1}{5.5/\sqrt{20}} = 1.14 \tag{3.7}$$

For $\alpha=0.05$ and the two-sided alternative $H_A : \mu \neq 3.1$, the critical values of t with $n-1$ degrees of freedom are $t_{0.05,19}=-2.09$ and $+2.09$. Since the observed value of the test statistic falls within the range of the critical values, we accept H_0 and conclude that there is not enough evidence to reject the null hypothesis. It is of course possible that we are making an error – specifically the Type II error of accepting a false hypothesis.

The p-value associated with the observed value of t is $2(.1325)=0.265$, and since this is greater than 0.05 it is consistent with the fact that we have not rejected the null hypothesis. In situations where the null hypothesis is true, we would expect more extreme values of t about 26.5% of the time. We would reject the null hypothesis only if our observed value of t was so extreme that it would be expected less than 5% of the time.

A 95% confidence interval for the mean is

$$\bar{x} \pm t_{0.05,19} s/\sqrt{n} = 4.5 \pm 2.09(5.5)/\sqrt{20} = 4.5 \pm 2.57 = (1.93, 7.07) \qquad (3.8)$$

Note that this interval includes the hypothesized value of 3.1.

3.3 One-Sample Tests for Proportions

When we are interested in whether a proportion, rather than a mean, is different from some hypothesized value, we need to know the sampling distribution of proportions when the null hypothesis is true. Suppose that the true proportion in a population is equal to ρ_0. Then the sampling distribution of proportions is normal, with mean ρ_0 and standard deviation $\sqrt{\rho_0(1-\rho_0)}$. This means that we may test hypotheses of the form

$$H_0 : \rho = \rho_0 \qquad (3.9)$$

by using a z-statistic of the form

$$z = \frac{p - \rho_0}{\sqrt{\rho_0(1-\rho_0)/n}} \qquad (3.10)$$

Even though we don't know the true value of ρ_0, when calculating z we simply use the hypothesized value ρ_0.

Note that the form of the z-statistic is always the same – the numerator is equal to the observed sample value minus the hypothesized value, and the denominator is equal to the standard deviation of the sampling distribution when H_0 is true. The z-statistic tells us how many standard deviations the observed value is away from the hypothesized value.

Note also that we can find a $100(1-\alpha)$% confidence interval around the sample proportion, as follows:

$$p - Z_\alpha \sqrt{\frac{p(1-p)}{n}} \leq \rho \leq p + Z_\alpha \sqrt{\frac{p(1-p)}{n}} \qquad (3.11)$$

3.3.1 *Illustration*

Suppose we are interested in knowing whether the proportion of households in an area who own two cars differs from the citywide figure of 0.2. We survey $n = 50$ households, and find $p = 16/50 = 0.32$. To test $H_0 : \rho = 0.2$ against the two-sided alternative $H_A : \rho \neq 0.2$ with $\alpha = 0.05$, we find

$$z = \frac{.32 - .2}{\sqrt{.2(.8)/50}} = 2.12 \qquad (3.12)$$

Since the observed value of z falls outside the range of the two critical values, $Z_{0.05} = \pm 1.96$, we reject the null hypothesis and conclude that the proportion of households that own two cars in this neighborhood is significantly higher than the citywide proportion.

The p-value is found by using the z-table (Table A.2) to determine the likelihood of a more extreme z-value than the one observed. The table reveals that $p(z > 2.12) = 0.017$. Since the probability of a z-value less than -2.12 is also 0.017, the probability of getting a statistic more extreme than the one observed is $2(0.017) = 0.034$. Note that the p-value is less than 0.05, and this is consistent with rejecting H_0.

3.4 Two-Sample Tests

3.4.1 Two-Sample t-Tests for the Mean

Often a sample mean is compared with another sample mean, rather than with some known population value. In this case the t-test is appropriate, and the form of the t-test depends upon whether the variances of the two samples can be assumed equal. The assumption of equal variances is known as *homoscedasticity*. If the variances can be assumed equal, the t-statistic is

$$t = \frac{\bar{x}_1 - \bar{x}_2}{s_p \sqrt{(1/n_1) + (1/n_2)}} \tag{3.13}$$

where x_1 and x_2 are the observed means of the two samples, n_1 and n_2 are the observed sample sizes, and the pooled estimate of the standard deviation, s_p, is equal to

$$s_p = \sqrt{\frac{(n_1 - 1)\, s_1^2 + (n_2 - 1)\, s_2^2}{n_1 + n_2 - 2}} \tag{3.14}$$

Here s_1^2 and s_2^2 represent the observed variances of samples 1 and 2, respectively. The number of degrees of freedom associated with this test statistic is $n_1 + n_2 - 2$, since the total sample size is effectively reduced by two due to the estimation of two means. Note that an alternative way of writing this t-statistic is

$$t = \frac{\bar{x}_1 - \bar{x}_2}{\sqrt{(s_p^2/n_1) + (s_p^2/n_2)}} \tag{3.15}$$

If it cannot be assumed that the two samples have equal variances, then the appropriate t-statistic is

$$t = \frac{\bar{x}_1 - \bar{x}_2}{\sqrt{(s_1^2/n_1) + (s_2^2/n_2)}} \tag{3.16}$$

In this case the number of degrees of freedom is more difficult to calculate – it is equal to (Sachs 1984)

$$\mathrm{df} = \frac{\left[(s_1^2/n_1) + (s_2^2/n_2)\right]^2}{\left[s_1^4/n_1^2(n_1 - 1)\right] + \left[s_2^4/n_2^2(n_2 - 1)\right]} \tag{3.17}$$

Some statistics texts suggest that the degrees of freedom be simply taken as the minimum of the two quantities $(n_1 - 1,\ n_2 - 1)$. Sachs notes that the degrees of freedom as determined from the expression above will always be between $\min(n_1 - 1,\ n_2 - 1)$ and $n_1 + n_2 - 2$. Therefore, taking the degrees of freedom to be equal to $\min(n_1 - 1,\ n_2 - 1)$ will be conservative, in the sense that the probability of committing the Type I error of rejecting a true hypothesis will be less than the nominal value of α. In other words, the critical value will be more extreme, and one will be less likely to reject the null hypothesis.

One may use the F-test to determine whether the assumption of equal variances is justified. Under the null hypothesis of equal variances, the test statistic

$$F = \frac{s_1^2}{s_2^2} \tag{3.18}$$

has an F-distribution, with $n_1 - 1$ and $n_2 - 1$ degrees of freedom in the numerator and denominator, respectively.

As is the case with the one-sample t-test, we must also assume that the populations from which the samples are taken are themselves normally distributed. Also, a z-test may be used instead of a t-test when sample sizes are large.

Illustration. Suppose we are interested in knowing whether differences in recreational behavior exist between the central city and suburban regions of a metropolitan area. In particular, suppose we are interested in swimming frequencies. Before collecting the data, we have no prior hypothesis regarding whether one region's individuals will have higher frequencies than the other, and so we will use a two-tailed test. The null and alternative hypotheses may be stated as

$$H_0 : \mu_{\mathrm{cc}} = \mu_{\mathrm{sub}} \qquad H_A : \mu_{\mathrm{cc}} \neq \mu_{\mathrm{sub}} \tag{3.19}$$

We collect the hypothetical data shown in Table 3.1, based on a random sample of eight residents in each region. Because the sample size is small, we will use a t-test; if the sample size were larger (say 30 or so residents from each region), we would use a z-test. An examination of the sample means reveals that the annual frequency is higher in suburban locations. Is this difference "significant," or might it have arisen by chance? With respect to the latter possibility, it could be the case that the observed difference is attributable to sampling fluctuation – if we took another sample of eight residents from each region, we might not find such a large difference. To proceed with the two-sample t-test, we should first decide whether the population variances

Table 3.1 **Annual swimming frequencies for eight central city and eight suburban residents**

	Annual swimming frequencies	
	Central city	Suburbs
	38	58
	42	66
	50	80
	57	62
	80	73
	70	39
	32	73
	20	58
Mean	48.63	63.63
Standard deviation	19.88	12.66

are equal. Using the F-test with $\alpha = 0.05$, we have

$$F = \frac{s_1^2}{s_2^2} = \frac{19.88^2}{12.66^2} = 2.47 < F_{\text{crit}} = F_{0.05,7,7} = 3.79 \tag{3.20}$$

The critical value of $F = 3.79$ comes from the F-table (Table A.4 in Appendix A). We therefore accept the assumption of equal variances. The sampling distribution has the form of a t-distribution with $n_1 + n_2 - 2 = 14$ degrees of freedom. Using the t-table (Table A.3a) with 14 degrees of freedom and a two-tailed test with $\alpha = 0.05$ implies that the critical values of t are -2.14 and 2.14. Again, the sampling distribution may be thought of as the frequency distribution resulting from many replications of the experiment (where "experiment" is defined here as surveying eight residents from each region) under the condition that the null hypothesis is true. If the null hypothesis of no difference between central cities and suburbs is true, then 5% of the time we can expect the observed t-value to be either less than -2.14 or greater than 2.14. In those 5% of the cases we would be making a Type I error by rejecting a true hypothesis.

Using Equation 3.13, we find that

$$t = \frac{63.63 - 48.63}{16.67\sqrt{\frac{1}{8} + \frac{1}{8}}} = 1.8 \tag{3.21}$$

Since our observed value of t is less than the critical value of 2.14, we fail to reject the null hypothesis.

We can also find the likelihood of obtaining a result that is more extreme than the one we observed, assuming that the null hypothesis is true (i.e., the p-value). Figure 3.5 shows that the p-value associated with the test is equal to $.0467 + .0467 = .0934$. This is found by using a t-table with 14 degrees of freedom and finding the area to the left of -1.8 and to the right of 1.8 (see Table A.3b). The p-value tells us the likelihood of getting a more extreme result than

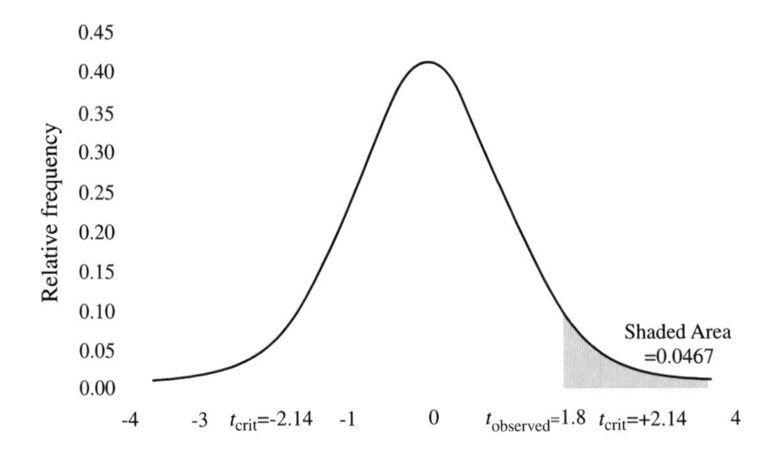

Figure 3.5 ***t*-distribution with 14 degrees of freedom**

the one we observed, if H_0 is true. Low p-values (i.e., lower than α) are coincident with rejection of H_0, since they imply that it would be quite unlikely to get a more extreme t-statistic than the one observed if H_0 were true. In our case, we have failed to reject the null hypothesis. The p-value gives us added information about precisely how unlikely our results are under the null hypothesis – if H_0 is true, we would expect a t-statistic with absolute value 1.8 or greater about 9.34% of the time. So we have observed a t-value that would be a bit unusual if the null hypothesis were true, but it is not unusual enough to reject the null hypothesis.

It is interesting to see what would have happened had we not assumed the variances of the two columns of data were equal. In that case, we would have had

$$t = \frac{63.63 - 48.63}{\sqrt{(19.88^2/8) + (12.66^2/8)}} = 1.8 \tag{3.22}$$

The observed t-value is the same, but the conservative number of degrees of freedom associated with the t-distribution is now $\min(8-1, 8-1) = 7$. Consulting a t-table reveals that the critical values are now -2.36 and 2.36. Since $1.8 < 2.36$, we again fail to reject the null hypothesis. Although the conclusion is the same, note that the p-value of $.0574 + .0574 = .1148$ is larger than it was before. A larger p-value and an observed t-value farther from its critical value than it was before imply that we do not come as close to rejecting the null hypothesis as we did when we assumed equal variances. When the variances are not assumed equal, it is more difficult to reject the null hypothesis. This illustration emphasizes the desirability of using the homoscedasticity assumption. In fact, we could have concocted a more interesting example where the observed value of t was equal to 2.2. Then we would have rejected H_0 under the first approach, assuming homoscedasticity, since $2.2 > 2.14$. We would have accepted H_0 under the second approach (assuming unequal variances), since $2.2 < 2.36$.

3.4.2 Two-Sample Tests for Proportions

When estimates of proportions are made from two samples taken from two identical populations, the distribution of differences in proportions is normal, with mean 0 and standard deviation equal to

$$\hat{\sigma}_p = \sqrt{\frac{p(1-p)}{n_1} + \frac{p(1-p)}{n_2}} \tag{3.23}$$

where p is the pooled estimate of the true proportion:

$$p = \frac{n_1 p_2 + n_2 p_2}{n_1 + n_2} \tag{3.24}$$

This means that we can test null hypotheses of the form

$$H_0 : \rho_1 - \rho_2 = 0 \tag{3.25}$$

using a z-statistic

$$z = \frac{(p_1 - p_2) - (\rho_1 - \rho_2)}{\hat{\sigma}_{p_1 - p_2}} = \frac{(p_1 - p_2)}{\hat{\sigma}_{p_1 - p_2}} \tag{3.26}$$

Illustration. We are interested in knowing whether two communities have identical proportions of people who use mass transit. We expect that community A has a higher percentage of transit users than community B. The null and alternative hypotheses are:

$$\left.\begin{array}{lll} H_0 : \rho_A - \rho_B = 0, & \text{or} & H_0 : \rho_A = \rho_B \\ H_A : \rho_A - \rho_B > 0, & \text{or} & H_A : \rho_A > \rho_B \end{array}\right\} \tag{3.27}$$

We collect the following sample data:

$$\left.\begin{array}{ll} p_1 = 0.3, & n_1 = 39 \\ p_2 = 0.2, & n_2 = 50 \end{array}\right\} \tag{3.28}$$

The pooled estimate of the proportion is

$$p = \frac{0.3(39) + 0.2(50)}{39 + 50} = 0.244 \tag{3.29}$$

The z-statistic is

$$z = \frac{0.3 - 0.2}{\sqrt{(0.244/39) + (0.244/50)}} = 0.95 \tag{3.30}$$

With $\alpha = 0.05$, z_{crit} for this one-sided test is 1.645, and so we accept the null hypothesis and conclude that there is no difference between the two communities. The *p*-value for this example is 0.17, since this corresponds to the area under the standard normal curve that is more extreme than the observed *z*-value. Note the fact that the *p*-value is greater than 0.05 is consistent with accepting the null hypothesis.

3.5 Distributions of the Variable and the Test Statistic

A key distinction exists between the distribution of the variable of interest and the sampling distribution of the test statistic. This distinction is often not fully appreciated. Suppose that the distribution of distances traveled by park-goers from their residences to the park is governed by the "friction of distance" effect, irrespective of the weather. This effect is widely observed in many types of spatial interaction, where the distribution of trip lengths is characterized by many short trips and relatively fewer longer ones (see Figure 3.6). If we want to test the null hypothesis that the mean trip distance is the same on rainy days as it is on sunny days, we would take two samples. We might expect that both the sunny-day trip length distribution and the rainy-day trip length distribution would have shapes that are similar to the exponential distribution. To test the null hypothesis, we must compare the observed difference in mean distances with the sampling distribution of differences, derived by assuming H_0 to be true. The latter may be thought of as the histogram of differences when many samples are used to calculate many differences in means, when H_0 is true. We know from the central limit theorem that the means of variables are normally distributed (given a large enough sample size), even when the underlying variables themselves are not normally distributed. We also know that the

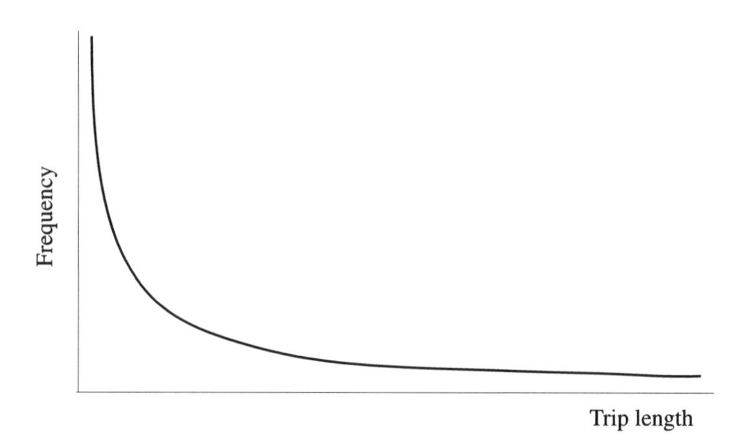

Figure 3.6 **Distribution of trip lengths**

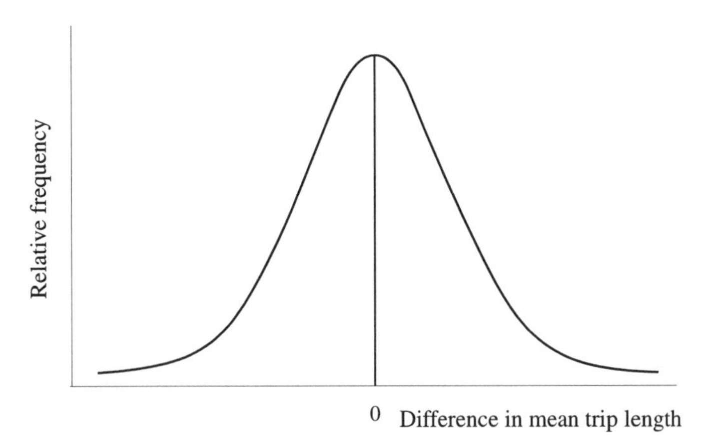

Figure 3.7 **Distribution of differences in mean trip length**

difference of two normal variables is normally distributed. Hence the two-sample test makes use of the fact that the sampling distribution of differences in means is normal (Figure 3.7). The important point is that there are two distributions to keep in mind – the distribution of the underlying variable (in this case exponential), and the distribution of the test statistic (in this case normal).

3.6 Spatial Data and the Implications of Nonindependence

One of the assumptions of the one- and two-sample t-tests is that observations are independent. This means that the observed value of one observation is not affected by the value of another observation. At first glance, this assumption sounds innocent enough, and it is tempting to simply ignore it and hope that it is satisfied. However, spatial data are often *not* independent; the value of one observation is very likely to be influenced by the value of another observation. In the swimming example, two individuals chosen at random in the central city are more likely to have similar responses than two individuals chosen at random from the suburbs. This could be because the accessibility of swimming pools is similar for them; the closer the two chosen individuals live together, the more similar is their distance to pools, and this would tend to make their swimming frequencies similar to one another. The closer two individuals live together, the more similar their incomes and lifestyles tend to be. This too would tend to cause similar swimming frequencies.

What are the consequences of a lack of independence among the observations? Because observations that are located near one another in space often exhibit similar values on variables, the effect is to reduce the effective sample size. Instead of n observations, the sample effectively contains information on less than n individuals. To take an extreme case, suppose that two individuals lived next door to one another, and thirty miles from the nearest pool. If we survey both of them, they are both likely to indicate that their swimming frequency was either zero or some very small number. The information contained in these two responses is essentially equivalent to the information contained in one response.

The implication of this is that when we carry out a two-sample t-test on observations that do not exhibit independence, we should really be using a critical value of t that is based on a smaller number of degrees of freedom than n. This in turn means that the critical value of t should be larger than the one that we use when we assume independence. A larger critical value of t means that it would be more difficult to reject the null hypothesis, and also that we are rejecting too many null hypotheses if we incorrectly assume independence. Thus there is a tendency to find significant results when in fact there are no significant differences in the underlying means of the two populations. The "apparent" differences between the two samples can, instead, be attributed to the fact that each sample contains observations that are similar to each other because of spatial dependence. Cliff and Ord (1975) give some examples of this, and supply the correct critical values of t that one should use, given a specified level of dependence.

When data are independent and the variance is σ^2, we have seen that a 95% confidence interval for the mean, μ, is $(\bar{x} - 1.96\sigma/\sqrt{n}, \bar{x} + 1.96\sigma/\sqrt{n})$. Following Cressie (1993), suppose that we collected $n = 10$ observations. For example, we might collect air quality data systematically along a transect (Figure 3.8). Let us choose x_1 from a normal distribution with mean μ and variance σ^2. Then, instead of choosing x_2 from a normal distribution with mean μ and variance σ^2, choose x_2 as

$$x_2 = \rho x_1 + \varepsilon \tag{3.31}$$

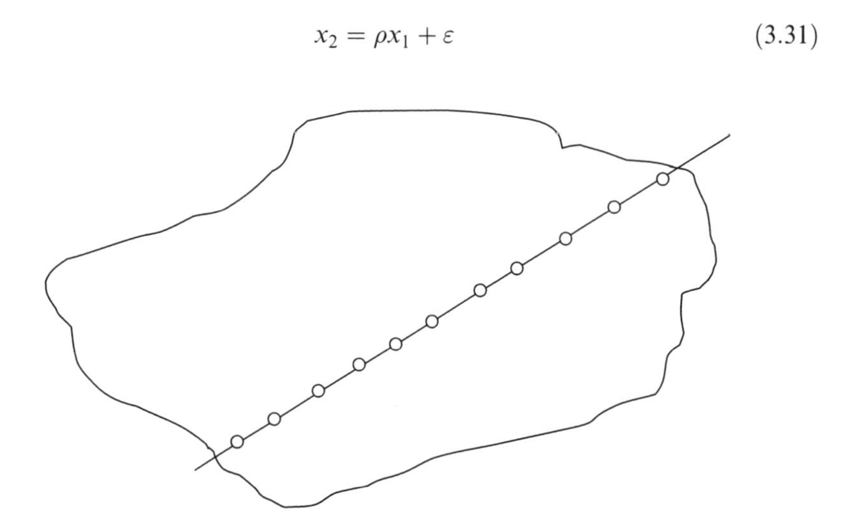

Figure 3.8 **Systematic collection of data along a transect**

where ε comes from a normal distribution with mean 0 and variance $\sigma^2(1-\rho^2)$, and ρ is a constant between 0 and 1 indicating the amount of dependence (with $\rho=0$ implying independence and $\rho=1$ implying a perfect dependence, so that $x_2=x_1$). Cressie indicates that when successive points are chosen as in 3.31, the variance of the mean is equal to σ^2/n only when data are independent ($\rho=0$); more generally it is equal to

$$\sigma_{\bar{x}}^2=\frac{\sigma^2}{n}\left[1+\frac{2\rho(n-1)}{n(1-\rho)}-\frac{2\rho^2(1-\rho^{n-1})}{n(1-\rho)^2}\right] \tag{3.32}$$

Cressie gives an example for $n=10$ and $\rho=0.26$; in this case $\sigma_{\bar{x}}^2=\left(\sigma^2/10\right)\times[1.608]$, implying that a 95% two-sided confidence interval for μ is $(\bar{x}-2.458\sigma/\sqrt{n},\ \bar{x}+2.458\sigma/\sqrt{n})$. It is important to realize that this is *wider* than the confidence interval that results from assuming independence.

If we write the variance of the mean as $\sigma_{\bar{x}}^2=\left(\sigma^2/n\right)[f]$, where f is the inflation factor induced by the lack of independence, we can also write

$$\sigma_{\bar{x}}^2=\frac{\sigma^2 f}{n}=\frac{\sigma^2}{n'} \tag{3.33}$$

where $n'=n/f$ is the effective number of independent observations. With $n=10$ and $\rho=0.26$, $f=1.608$ and $n'=10/1.608=6.2$; this means that our 10 *dependent* observations are equivalent to a situation where we have $n'=6.2$ independent observations.

3.7 Sampling

The statistical methods discussed throughout this book rely upon sampling from some larger population. The population may be thought of as the collection of all elements or individuals that are the object of our interest. The list of all elements in the population is referred to as the *sampling frame*. Sampling frames may consist of spatial elements – for example, all of the census tracts in a city. We may be interested in the commuting times of all individuals in a community, or in the migration distances of all people who have moved during the past year. It is important to have a clear definition of this population, since this is the group about which we are making inferences. The inferences are made using information collected from a sample.

There are many ways to sample from a population. Perhaps the simplest sampling method is *random sampling*, where each of the elements has an equal probability of being selected from the population into the sample. For example, suppose we wish to take a random sample of size $n=4$ from a population of size $N=20$. (A common convention is to use upper case "N" to denote population size and lower case "n" to denote sample size.) Choose a random number from 1 to 20. Then select another random number from 1 to 20. If it is the same as the previous random number, discard it and choose another. Repeat this until four distinct random numbers, representing elements of the sampling frame, have been chosen. To illustrate, we will use the first two digits of the five-digit random numbers from Table A.1 in Appendix A. Beginning at the upper-left of the table and proceeding down the column, the first two-digit number in the range 01–20 is 17. To complete our sample of $n=4$, we proceed down the column and choose the next three numbers in this range – they are 04, 03, and 07.

Choosing a *systematic sample* of size n begins by selecting an observation at random from among the first $[N/n]$ elements, where the square brackets indicate that the integer part of N/n is to be taken. Thus if N/n is not an integer, one just uses the integer part of N/n. Call the label of this randomly chosen element k. The elements of the sampling frame that are in the sample are $k+i[N/n]$, $i=0, 1, \ldots, n-1$. With $N=20$ and $n=4$, $k=N/n=5$. Suppose, from among the first five elements, we choose element $k=2$ at random. The elements in the sample are 2, $2+5=7$, $2+10=12$, and $2+15=17$. Note that it was necessary to choose only one random number.

When it is known beforehand that there is likely to be variation across certain subgroups of the population, the sampling frame may be *stratified* before sampling. For example, suppose that our $N=20$ individuals can be divided into two groups – $N_m=15$ men and $N_w=5$ women. A *proportional, stratified* sampling of individuals is achieved by making the sample proportions in each stratum equal. Thus we could choose $n_m=3$ men randomly from among the group of $N_m=15$, and $n_w=1$ woman randomly from the group of $N_w=5$ women. For both men and women, the sampling proportion is 1/5.

When the sampled size of the stratum is small, it may be advantageous to obtain a *disproportional* random sample, where the small group is oversampled. In the case above, using $n_m=2$ and $n_w=2$ would result in unequal sample proportions, since $n_m/N_m=2/15$ for men and $n_w/N_w=2/5$ for women.

3.7.1 Spatial Sampling

When the sampling frame consists of all of the points located in a geographical region of interest, there are again several alternative sampling methods.

A *random spatial sample* consists of locations obtained by choosing x-coordinates and y-coordinates at random. If the region is a non-rectangular shape, x- and y-coordinates may be chosen by selecting them at random from the ranges $(x_{\min}, x_{\max})$ and $(y_{\min}, y_{\max})$. If the pair of coordinates happens to correspond to a location outside of the study region, the point is simply discarded.

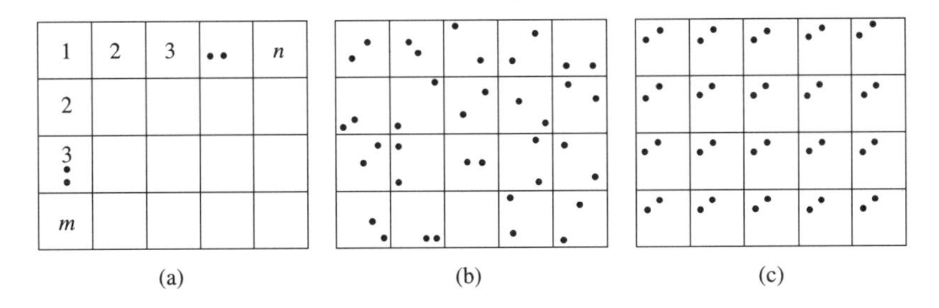

Figure 3.9 **Examples of spatial sampling: (a) study region stratified into subregions; (b) stratified spatial sampling; (c) systematic spatial sampling**

To ensure adequate coverage of the study area, the study region may be broken into a number of mutually exclusive and collectively exhaustive strata. Figure 3.9(a) divides a study region into a set of $s = mn$ strata. A *stratified spatial sample* of size mnp is obtained by taking a random sample of size p within each of the mn strata (Figure 3.9b). A *systematic spatial sample* of size mnp is obtained by (i) taking a random sample of size p within any individual stratum, and then (ii) using the sample spatial configuration of those p points within that stratum within the other strata (Figure 3.9c).

The question of which sampling scheme is "best" depends upon the spatial characteristics of variability in the data. In particular, because values of variables at one location tend to be strongly associated with values at nearby locations, random spatial samples can provide redundant information when sample locations are close to one another. Consequently, stratified and systematic random sampling tend to provide better estimates of the variable's mean value. Thus if one were to repeat the sampling many times, the variability associated with the means calculated using systematic or stratified sampling would be less than that found with random spatial sampling. Haining (1990a) discusses this in more detail, and gives references that suggest that systematic random sampling is often slightly better than stratified random sampling.

3.8 Two-Sample *t*-Tests in *SPSS for Windows 9.0*

3.8.1 Data Entry

Suppose we wish to enter the data from Table 3.1 into *SPSS* and conduct a two-sample *t*-test of the null hypothesis that the mean annual swimming frequency among residents of the central city is equal to the mean annual swimming frequency among residents of the suburbs.

We begin by entering the data. In *SPSS*, this entails entering all of the swimming frequencies into one column. Another column contains a numeric value indicating which region the corresponding swimming frequency belongs to. For our two-region example, we would have

Swim	Location
38	1
42	1
50	1
57	1
80	1
70	1
32	1
20	1
58	2
66	2
80	2
62	2
73	2
39	2
73	2
58	2

The variable names "Swim" and "Location" are defined by right-clicking at the head of each column on the heading "var" that appears in the *SPSS* data editor. Then, under Define Variable, variable names may be assigned.

Note that here the first eight rows correspond to the data from the central city; location 1 refers to the central city. Similarly, the last eight rows contain the value "2" in the second column, and these correspond to the observations from the suburbs. In general, if there are n_1 observations on one variable and n_2 observations on the other, then there will be $n_1 + n_2$ rows and 2 columns once data have been entered into *SPSS*.

3.8.2 Running the t*-Test*

To run the analysis within *SPSS*, click on Analyze (Statistics in earlier versions of *SPSS for Windows*), then on Compare Means, and then on Independent Samples *t*-test. A box will open, and the variable Swim should then be highlighted and moved to the Test Variable box via the arrow tab. The variable Location is moved to the box headed Grouping Variable (since we are testing the variable Swim for differences by Location). Under the Grouping Variable box, click on Define Groups, and enter 1 for Group 1 and 2 for Group 2; these are the numeric values that *SPSS* will use from the second column of data to distinguish between groups. Then click Continue. Under Options, the percentage associated with confidence interval may be assigned if desired (the default is 95%). Finally, click OK.

An example of the output from a two-sample *t*-test is shown in Table 3.2, which depicts the results of the test of equality of swimming frequencies in central city and suburbs using *SPSS 9.0 for Windows*.

Table 3.2 **Results of two sample *t*-test**

Group Statistics

	LOCATION	N	Mean	Std. Deviation	Std. Error Mean
Swimfreq	1.00	8	48.6250	19.8778	7.0278
	2.00	8	63.6250	12.6597	4.4759

Independent Samples Test

		Levene's Test for Equality of Variances		t-test for Equality of Means						
									95% Confidence Interval of the Difference	
		F	Sig.	t	df	Sig. (2-tailed)	Mean Difference	Std. Error Difference	Lower	Upper
SWIMFREQ	Equal variance assumed	1.776	.204	−1.800	14	.093	−15.0000	8.3321	−32.8706	2.8076
	Equal variance not assumed			−1.800	11.876	.097	−15.0000	8.3321	−33.1751	3.1751

First, the swimming frequencies in each region are summarized; location 1 (central city) has a mean response of 48.625 days and a standard deviation of 19.8778, while those in the suburbs apparently swim more often – the responses there have a mean of 63.625 and a standard deviation of 12.6597.

Below this are the results of the analysis. First, note that there is a test of the assumption that the variances of the two groups are indeed equal. This test, Levene's test, is based upon an *F*-statistic. The key piece of output is the column headed "Sig.", since this tells us whether to accept or reject the null hypothesis that the two variances are equal. Since this value (which is also known as a *p*-value) is greater than 0.05, we can accept the null hypothesis, and conclude that the variances may be assumed equal.

The results of the *t*-test are given for both instances – one where the variances are assumed equal, and one where they are not. In both cases, the *t*-statistic is 1.8, and in both cases we accept the null hypothesis since the "Sig." column indicates a value higher than 0.05. Note that, when equal variances are assumed, we come slightly closer to rejecting the null hypothesis (the *p*-value in that case is 0.093, compared with 0.097 when the variances are not assumed equal). The *p*-values differ despite identical *t*-statistics because the degrees of freedom differ.

Finally, note that the 95% confidence intervals include zero, indicating that the true difference between city and suburbs could be zero.

Exercises

1. A political geographer is interested in the spatial voting pattern during a recent presidential election involving two candidates, A and B. She suspects that university professors were more likely than the general population to vote for candidate A. She takes a random sample of 45 professors in the state, and finds that 20 voted for the candidate A. Is there sufficient evidence to support her hypothesis? The statewide percentage of the population voting for candidate A was 0.38. What is the *p*-value?

2. A survey of the white and nonwhite population in a local area reveals the following annual trip frequencies to the nearest state park:

$$\begin{array}{lll} \bar{x}_1 = 4.1, & s_1^2 = 14.3, & n_1 = 20 \\ \bar{x}_2 = 3.1, & s_2^2 = 12.0, & n_2 = 16 \end{array}$$

(a) Assume that the variances are equal, and test the null hypothesis that there is no difference between the park-going frequencies of whites and nonwhites.
(b) Repeat the exercise, assuming that the variances are unequal.
(c) Find the *p*-value associated with the tests in parts (a) and (b).
(d) Find a 95% confidence interval for the difference in means.
(e) Repeat parts (a)–(d), assuming sample sizes of $n_1 = 24$ and $n_2 = 12$.

3. Test the hypothesis that two communities have equal support for a political candidate using the following data:

$$\text{Community A:} \quad p = 0.33, \quad n_A = 54$$
$$\text{Community B:} \quad p = 0.18, \quad n_B = 38$$

In addition to testing the hypothesis, find the p-value.

4. A researcher suspects that the level of a particular stream's pollutant is higher than the allowable limit of 4.2 mg/l. A sample of $n = 17$ reveals a mean pollutant level of $\bar{x} = 6.4$ mg/l, with a standard deviation of 4.4. Is there sufficient evidence that the stream's pollutant level exceeds the allowable limit? What is the p-value?

5. Information is collected by a researcher from 14 individuals on their use of rapid transit. Seven individuals were from suburb A and seven were from suburb B. The following data are the number of times per year the individual used rapid transit:

Individual	Suburb A	Suburb B	Pooled data
1	5	67	
2	12	56	
3	14	44	
4	54	22	
5	34	16	
6	14	61	
7	23	37	
Mean	22.29	43.28	32.79
Std. dev.	16.67	19.47	20.57

Do the suburbs differ with respect to the mean number of rapid transit trips taken by individuals? Use the two-sample t-test, assuming the variances are equal. Give the critical value of t, recalling that the number of degrees of freedom is equal to $n_1 + n_2 - 2$. What is the p-value associated with this test?

6. The contour lines of the map below represent elevation.

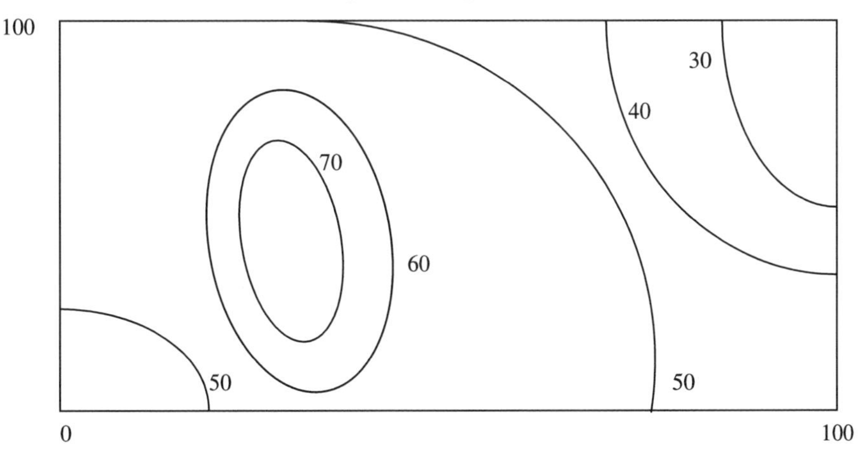

(a) Take a random spatial sample of $n = 18$ points and estimate the mean elevation of the study area.
(b) Divide the study region into a set of $3 \times 3 = 9$ strata of equal size. Take a stratified sample of size 18 by randomly choosing two points from within each stratum. Estimate the mean.
(c) Using the same $3 \times 3 = 9$ strata from (b), choose a systematic random sample by first randomly selecting two points from within any individual stratum. Then use the configuration of points within that stratum to select points within the other strata (see Figure 3.9c). Estimate the mean elevation from the resulting 18 points.

Note: If answers from the entire class are pooled together, it will usually (but not always!) be the case that the means found in part (a) will display greater variability than those found in parts (b) and (c).

7.
(a) A two-tailed test of a one-sample hypothesis of a mean yields a test statistic of $z = 1.47$. What is the p-value?
(b) A one-tailed test of a two-sample hypothesis involving the difference of sample means yields $t = 1.85$, with 12 degrees of freedom. What is the p-value?

4 Analysis of Variance

LEARNING OBJECTIVES

- Comparison of means in three or more samples
- Analysis of underlying assumptions
- Introduction of alternative tests for cases where assumptions are not reasonable

4.1 Introduction

The two-sample difference of means test may be generalized to the case of more than two samples. In this case, we wish to test the null hypothesis that the population means from a set of $k > 2$ are all equal:

$$H_0 : \mu_1 = \mu_2 \cdots = \mu_k \tag{4.1}$$

Such hypotheses may concern variation in means over time or space. For example, we may wish to know whether traffic counts vary by month, or whether the number of weekly shopping trips made by households varies among the central city, suburban, and rural portions of a county.

Data for such problems are typically given in a table such as Table 4.1, with the categories constituting the columns. Note that the notation in the table is a bit different – "pluses" are used to indicate means, so that X_{+1} designates the mean of column one, and X_{++} denotes the mean of all observations, summed over rows and columns.

Analysis of variance (ANOVA) represents an extension of the two-sample *t*-test for differences of means. It involves the introduction of some new ideas, though the underlying assumptions of the test are similar to those used in the two-sample *t*-test.

The assumptions of analysis of variance may be stated as follows:

(1) Observations between and within samples are random and independent.
(2) The observations in each category are normally distributed.
(3) The population variances are assumed equal:

$$\sigma_1^2 = \sigma_2^2 \cdots \sigma_k^2 = \sigma^2 \tag{4.2}$$

The assumed equality of variances is referred to as the assumption of *homoscedasticity* (sometimes written as *homoskedasticity*). Though the analysis of

Table 4.1 **Arrangement of data for analysis of variance**

	Category 1	Category 2	... Category k
Obs. 1	X_{11}	X_{12}	X_{1k}
Obs. 2	X_{21}	X_{22}	X_{2k}
Obs. 3	X_{31}	X_{32}	X_{3k}
⋮			
Obs. i	X_{i1}	X_{i2}	X_{ik}
No. of obs.	n_1	n_2	n_k
Mean	X_{+1}	X_{+2}	X_{+k}
Standard deviation	s_1	s_2	s_k
Overall mean: X_{++}			

variance test is one that tests for the equality of group means, the test itself is carried out using two independent estimates of the common variance σ^2. One estimate of the variance is a pooled estimate of the within-group variances. The other estimate of the variance is a between-group variance.

The idea behind the test is to compare the variation within columns to the variation between column means. If the variation between group means is much greater than the variation within columns, we will be inclined to reject the null hypothesis. If, however, the variation between group means is not very large relative to the variation within columns, this suggests that any differences in group means may be due to sampling fluctuation, and hence we are more inclined to accept H_0. For example, in Table 4.1, there is variability within columns; different individuals within each subregion have differing levels of participation. There is also variability between columns; the sample means in each region are different. If the between-column variability is high relative to the within-column variability, we will reject the null hypothesis and conclude that the true column means are not equal.

To be more specific, we may define the total sum of squares as the sum of the squared deviations of all observations from the overall mean. This total sum of squares (TSS) may be partitioned into a "between sum of squares" (BSS) and a "within sum of squares" (WSS). The comparison of between-column variation to within-column variation leads to an F-statistic. The partitioning of the sum of squares is as follows:

$$\left.\begin{aligned} \text{TSS} &= \sum_i \sum_j (X_{ij} - X_{++})^2 \\ \text{BSS} &= \sum_j n_j (X_{+j} - X_{++})^2 \\ \text{WSS} &= \sum_i \sum_j (X_{ij} - X_{+j})^2 = \sum_j (n_j - 1)\, s_j^2 \end{aligned}\right\} \tag{4.3}$$

The F-statistic is

$$F = \frac{\text{BSS}/(k-1)}{\text{WSS}/(N-k)} \tag{4.4}$$

When the null hypothesis is true, this statistic has an F-distribution, with $k-1$ degrees of freedom and $N-k$ degrees of freedom associated with the numerator and denominator, respectively.

4.1.1 A Note on the Use of F-Tables

F-statistics are based on ratios. There are degrees of freedom associated with both the numerator and the denominator. F-tables are typically arranged so that the columns correspond to particular degrees of freedom associated with the numerator, and rows correspond to particular degrees of freedom associated with the denominator. Entries in the table give the critical F-values, and the entire table is associated with a given significance level α. Many texts give separate tables for $\alpha = 0.01$, 0.05, and 0.10; these are provided, for example, in Table A.4 in Appendix A. Because F-tables are displayed in this way, it is often difficult to state the p-value associated with the test. Stating the p-value would require a very complete set of F-tables for many values of α. Software for statistical analysis is often useful in this regard, since p-values are usually provided.

4.1.2 More on Sums of Squares

To see why the between sum of squares plus the within sum of squares add to the total sum of squares, recognize that the total sum of squares may be written as

$$\sum_i \sum_j (X_{ij} - X_{++})^2 = \sum_i \sum_j \left[(X_{ij} - X_{+j}) + \sum_i \sum_j (X_{+j} - X_{++}) \right]^2 \tag{4.5}$$

Expanding the square yields

$$\sum_i \sum_j (X_{ij} - X_{++})^2 = \sum_i \sum_j (X_{ij} - X_{+j})^2 + 2(X_{ij} - X_{+j})(X_{+j} - X_{++}) + (X_{+j} - X_{++})^2 \tag{4.6}$$

The middle term on the right-hand side is equal to zero, since the sum of deviations from a mean is equal to zero:

$$\sum_i (X_{ij} - X_{+j}) = 0 \tag{4.7}$$

The first term on the right-hand side is equal to the within sum of squares:

$$\text{WSS} = \sum_i \sum_j (X_{ij} - X_{+j})^2 \tag{4.8}$$

Similarly, the last term on the right-hand side is equal to the between sum of squares:

$$\text{BSS} = \sum_i \sum_j (X_{+j} - X_{++})^2 = \sum_j n_j (X_{+j} - X_{++})^2 \tag{4.9}$$

This demonstrates that the total sum of squares is equal to the between sum of squares plus the within sum of squares.

4.2 Illustrations

4.2.1 Hypothetical Swimming Frequency Data

Suppose we extend the previous example to include residents of the outlying rural region, using the data in Table 4.2. We formulate the null hypothesis of no difference in the mean annual swimming frequency between the three regions:

$$\mu_{\text{SUB}} = \mu_{\text{CC}} = \mu_{\text{R}} \tag{4.10}$$

With $\alpha = 0.05$, the critical value of F is equal to $F_{.05,2,21} = 3.47$. The observed F-statistic along with its components are given below:

$$\left.\begin{array}{l} \text{Total sum of squares: 6406.79} \\ \text{Between sum of squares: 1603.85} \\ \text{Within sum of squares: 4802.94} \\ F = \dfrac{1605.33/(3-1)}{4802.94/(24-3)} = 3.51 \end{array}\right\} \tag{4.11}$$

Since the observed F value exceeds the critical value of 3.47, the null hypothesis is rejected. How are the sums of squares most easily calculated? One way is to

Table 4.2 **Annual swimming frequencies for three regions**

	Annual swimming frequencies		
	Central city	Suburbs	Rural
	38	58	80
	42	66	70
	50	80	60
	57	62	55
	80	73	72
	70	39	73
	32	73	81
	20	58	50
Mean	48.63	63.63	67.63
Standard deviation	19.88	12.66	11.43
$X_{++} = 59.96$; $s^2 = 16.69$			

recognize that the total sum of squares is equal to the overall variance multiplied by $N-1$. Thus $6406.79=(16.69)^2(23)$. The within sum of squares is equal to the sum of the products of the group variances and (n_j-1). Thus $4802.94=(7*19.88^2)+(7*12.66^2)+(7*11.43^2)=7*(19.88^2+12.66^2+11.43^2)$. The between sum of squares is then derived as the difference between the total and within sum of squares: BSS = 6406.79 − 4802.94 = 1603.85.

4.2.2 Diurnal Variation in Precipitation

The effects of urban areas on temperature are well known – temperatures are generally higher in cities than in the surrounding countryside (this is known as the urban "heat island" effect). But what about the effects of urban areas on precipitation?

One possibility is that the particulate matter ejected by urban factories forms the condensation nuclei necessary for precipitation. If this is correct, one might expect to see diurnal variation in precipitation, since factories are typically idle on the weekends. If there is no lag, precipitation would be lightest on the weekends, and heaviest during the week.

I collected the data in Table 4.3 while I was an undergraduate, in conjunction with an assignment in my geography statistics class! For each day of the week, the data are lumped into six-month categories. One consequence of this lumping is to make the assumption of normality more plausible (since sums of variables from any type of distribution tend to be normally distributed).

A look at the data reveals that the largest amount of precipitation occurs on Fridays and Sundays, and the least on Tuesdays and Wednesdays. Perhaps there is a roughly two-day lag between the buildup of particulate matter during the week and the precipitation events.

Table 4.3 **Precipitation data for LaGuardia airport**

	Precipitation at LaGuardia airport (inches)						
Year	SAT	SUN	MON	TUE	WED	THUR	FRI
1971 II	2.30	6.84	4.47	3.40	0.94	1.71	8.30
1972 I	5.56	6.81	1.97	2.26	3.03	4.42	5.08
1972 II	5.31	1.50	1.74	3.00	5.89	4.16	2.88
1973 I	2.15	4.39	3.96	1.17	4.35	4.78	7.09
1973 II	1.71	4.12	2.87	0.79	3.90	3.11	5.68
1974 I	2.60	2.50	1.68	1.36	0.45	4.03	5.27
Mean	3.27	4.36	2.78	2.00	3.09	3.70	5.72
Std. dev.	1.70	2.18	1.20	1.06	2.08	1.12	1.86

Overall mean: 3.56; Overall std. dev.: 1.90

	Sum of squares	d.f.	Variance
Between:	51.97	6	8.663
Within:	96.34	35	2.753

$F=3.15$

$F_{0.05,6,35}=2.37$; $F_{0.01,6,35}=3.37$

The null hypothesis is that mean precipitation in each six-month period does not vary with day of the week. The results of the analysis of variance, shown in the table, reveal that the null hypothesis is rejected. In the exercises at the end of the chapter, you will be asked to repeat this analysis for Boston and Pittsburgh.

4.3 Analysis of Variance with Two Categories

The analysis of variance with two categories gives the same results as the two-sample t-test. To illustrate, consider once again the first two columns (central city and suburbs) of the swimming frequency data. Analysis of variance yields the following:

$$\left.\begin{aligned} &\text{BSS} = 900 \\ &\text{WSS} = 3888 \\ &F = \frac{(900/1)}{(3888/14)} = 3.24 \\ &F_{.05,1,14} = 4.60 \\ &F_{0.10,1,14} = 3.10 \end{aligned}\right\} \tag{4.12}$$

The null hypothesis of no difference is therefore rejected using $\alpha = 0.10$, and accepted using $\alpha = 0.05$. The p-value must be close to, but less than, 0.10. The result is in fact the same as that found in the last chapter using the t-test, under the assumption of equal variances.

When there are two categories, either the F-test or the t-test may be used.

4.4 Testing the Assumptions

Since the analysis of variance depends upon a number of assumptions, it is important to know whether these assumptions are satisfied. One way of testing the assumption of homoscedasticity is to use Levene's test. There are a number of ways to test normality. Two common methods are the Kolmogorov–Smirnov test and the Shapiro–Wilk test. Although a detailed discussion of all of these tests is beyond the scope of this text, most statistical software packages do provide these tests to allow researchers to test the underlying assumptions. Some additional details on each are given at the end of this chapter, in Section 4.8.

4.5 The Nonparametric Kruskal–Wallis Test

What do we do if the assumptions are not satisfied? We have at least two options. One is to proceed with the analysis of variance anyway, and "hope" that we get a valid answer. Fortunately, that is often not a bad way to proceed. The F-test is said to be relatively "robust" with respect to deviations from the assumptions of normality and homoscedasticity. This means that the results of the F-test may still be used effectively if the assumptions are at least "reasonably close" to being satisfied. If either (a) the assumptions are close to being satisfied, or (b) the F-statistic yields a "clear" conclusion (say, for example, a p-value much less than, say, 0.01, or greater than 0.20), the conclusion will generally be acceptable.

If the data deviate drastically from the assumptions, or if the p-value is close to the significance level, then an alternative test that does not rely on the assumptions might be considered. Tests that do not make assumptions regarding how the underlying data are distributed are called *nonparametric* tests. The nonparametric test for two or more categories is the Kruskal–Wallis test.

There is another set of circumstances in which the Kruskal–Wallis test is useful for testing hypotheses about a set of means – namely when only ranked (i.e., ordinal) data are available. In such situations, there is insufficient information to use ANOVA, which requires interval or ratio level data. (With interval and ratio data, the magnitude of the difference between the observations is meaningful.)

The application of the Kruskal–Wallis test begins by ranking the entire pooled set of N observations from lowest to highest. That is, the lowest observation is assigned a rank of 1, and the highest observation is assigned a rank of N. The idea behind the test is that, if the null hypothesis is true, then the sum of the ranks in each column should be about the same. Again, no assumptions about normality and homoscedasticity are required. The test statistic is

$$H = \left(\frac{12}{N(N+1)} \sum_{i=1}^{k} \frac{R_i^2}{n_i} \right) - 3(N+1) \tag{4.13}$$

where R_i is the sum of the ranks in category i, and n_i is the number of observations in category i. Under the null hypothesis of no difference in category means, the statistic H has a chi-square distribution with $k-1$ degrees of freedom. Table A.5 contains critical values for the chi-square distribution.

4.5.1 Illustration: Diurnal Variation in Precipitation

The LaGuardia airport precipitation data are ranked and displayed in Table 4.4. Also shown is the sum of the ranks for each column. Employing Equation 4.13 yields a value of $H = 13.17$. This is just slightly higher than the critical value of 12.59, and so the null hypothesis of no variation in precipitation by day of the week is rejected at the $\alpha = 0.05$ significance level. Note that the hypothesis would have been accepted using $\alpha = 0.01$. The p-value associated with the test is approximately 0.04, meaning that if the null hypothesis were true, a test statistic this high would be observed only 4% of the time.

Table 4.4 **Ranked precipitation data for LaGuardia airport**

	Ranks of observations (1 = lowest; 42 = highest)						
Year	SAT	SUN	MON	TUE	WED	THUR	FRI
1971 II	14	40	31	22	3	8	42
1972 I	36	39	11	13	20	30	33
1972 II	35	6	10	19	38	27	18
1973 I	12	29	24	4	28	32	41
1973 II	9	26	17	2	23	21	37
1974 I	16	15	7	5	1	25	34
SUM	122	155	100	65	113	143	205

Kruskal–Wallis statistic: $H = 13.17$
Critical value: $\chi^2_{0.05,6} = 12.59$; $\chi^2_{0.01,6} = 16.81$

The reader should compare this with the p-value associated with the ANOVA results. The ANOVA results yielded a p-value just slightly higher than 0.01 (we know this since the observed F-value is just slightly less than the critical F value of 3.37, using $\alpha = 0.01$). This result is a typical one – the Kruskal–Wallis test, though not relying on as many assumptions as the analysis of variance, is not as powerful. That is, it is harder to reject false hypotheses. Thus we would have rejected H_0 with ANOVA using, say, $\alpha = 0.02$ or above, whereas we would only have rejected H_0 using the Kruskal–Wallis test had we chosen $\alpha = 0.04$ or above.

4.5.2 More on the Kruskal–Wallis Test

If there are values for which the ranks are tied, an adjustment is made to the value of H. Suppose that we have $N = 10$ original observations, ranked from lowest to highest: 3.2, 4.1, 4.1, 4.6, 5.1, 5.2, 5.2, 5.2, 6.1, and 7.0. There are two sets of tied observations. When the data are assigned ranks, the tied values are each assigned the average rank. Thus the ranks of these ten observations are: 1, 2.5, 2.5, 4, 5, 7, 7, 7, 9, 10. In instances where tied ranks exist, the usual value of H is divided by the quantity

$$1 - \frac{\sum_i (t_i^3 - t_i)}{N^3 - N} \tag{4.14}$$

where t_i is the number of observations tied at a given rank, and the sum is over all sets of tied ranks. In our example containing ten observations, the adjustment is

$$1 - \frac{(2^3 - 2) + (3^3 - 3)}{10^3 - 10} = 1 - \frac{30}{990} = \frac{32}{33} \tag{4.15}$$

The effect of this adjustment is to make H bigger, and therefore to give the Kruskal–Wallis test slightly higher power, since it is then easier to reject false hypotheses.

The formula for the Kruskal–Wallis test appears rather mysterious, and the reader may wish to have a little more understanding of it. A glimpse of insight may be obtained by asking what the value of H would be if all of the observations were equal, and therefore all observations had the same rank. We will assume that there is an equal number of observations ($n = N/k$) per category. With N observations, the sum of the ranks is equal to the sum of the integers, from 1 to N:

$$S = \sum_{i=1}^{N} i = \frac{N(N+1)}{2} \tag{4.16}$$

(An historical aside that I was told when I was a boy: Gauss, at the age of 7, was punished by his schoolteacher. His punishment consisted of having to find the sum of all of the integers, from 1 to 100. Within a few minutes, he had figured out the formula above, and used it to find the answer (5050), rather than have to carry out the tedious task of actually summing all 100 numbers.)

Now, if all of the observations are tied, the average rank assigned to all N observations is S/N. Furthermore the sum of the ranks in each category i will be

$$R_i = n(S/N) = S/k = \frac{N(N+1)}{2k} \tag{4.17}$$

Using this in the first term on the right-hand side of Equation 4.13 yields

$$\frac{12}{N(N+1)} \sum_{i=1}^{k} \left(\frac{N(N+1)}{2k} \right)^2 \left(\frac{k}{N} \right) = \frac{12}{N(N+1)} \sum_{i=1}^{k} \frac{N(N+1)^2}{4k} = 3(N+1) \tag{4.18}$$

We have just shown that H is therefore equal to zero when all of the ranks are tied.

4.6 Contrasts

The analysis of variance, as a test for the equality of means, can sometimes leave the analyst with a sense of unfulfillment. In particular, if the null hypothesis is rejected, what have we learned? We've learned that there is significant evidence to conclude that the means are not equal, but we do not know *which* means differ from one another. We might look at the data and get a feel for which means seem high and which seem low, but it would be nice to have a way of testing to see whether particular combinations of categories had significantly different means. We may, for example, want to know, in an example involving five categories, whether the difference between categories 2 and 5 (that is, $\mu_2 - \mu_5$) differs significantly from zero. Differences that are of interest may involve more than two separate means. For instance, with the precipitation data, we may wish to contrast weekends with weekdays. In that case, we would want to contrast the mean of the first two categories (Saturday and Sunday) with the mean of the last five, weekday categories. This could be represented as

$$\frac{\mu_{\text{SAT}} + \mu_{\text{SUN}}}{2} - \frac{\mu_{\text{MON}} + \cdots + \mu_{\text{FRI}}}{5} \tag{4.19}$$

Scheffé (1959) described a formal procedure for contrasting sets of means with one another. A *contrast*, ψ, is defined as a combination of the means. Usually, one defines linear combinations, so

$$\psi = \sum_{i=1}^{k} c_i \mu_i \tag{4.20}$$

where the values of c_i are specified by the analyst in a manner that is consistent with the contrast of interest. In our first example, categories 2 and 5 would be contrasted with each other using the values $c_1 = 0$, $c_2 = 1$, $c_3 = 0$, $c_4 = 0$, and $c_5 = -1$. This choice of coefficients arises from writing

$$\mu_2 - \mu_5 = 0\,\mu_1 + 1\,\mu_2 + 0\,\mu_3 + 0\,\mu_4 + (-1)\,\mu_5 \tag{4.21}$$

In the precipitation example, the coefficients for the weekend days would each be equal to 1/2, and the coefficients for the weekdays would each be $-1/5$. Why? This combination arises from realizing that Equation 4.20 can be written as

$$\begin{aligned} \psi &= \frac{\mu_{\text{SAT}} + \mu_{\text{SUN}}}{2} - \frac{\mu_{\text{MON}} + \cdots + \mu_{\text{FRI}}}{5} \\ &= 1/2\,\mu_{\text{SAT}} + 1/2\,\mu_{\text{SUN}} - 1/5\,\mu_{\text{MON}} - \cdots - 1/5\,\mu_{\text{FRI}} \end{aligned} \tag{4.22}$$

Scheffé's contribution was to show that simultaneous confidence intervals at the $1 - \alpha$ level for all possible contrasts are given by

$$\psi - R\sigma_\psi \leq \psi \leq \psi + R\sigma_\psi \tag{4.23}$$

where

$$R = \sqrt{(k-1)\,F_{k-1,N-k,\alpha}} \tag{4.24}$$

and

$$\sigma_\psi^2 = \left(\frac{\text{WSS}}{N-k}\right) \sum_{j=1}^{k} \frac{c_j^2}{n_j} \tag{4.25}$$

If the original null hypothesis of no difference among the k means is rejected, then there is at least one contrast that is significantly different from zero. If we have one or more contrasts that we wish to test, we may do so using the inequality in (4.23). The null hypothesis that any given contrast is equal to zero is rejected if the confidence interval (4.23) does not contain zero. If the original null hypothesis of no difference among the k means is *not* rejected, then there are no contrasts that will be found significant.

4.6.1 A Priori *Contrasts*

The contrasts just described are *a posteriori* or *post hoc* contrasts, since they occur after the analysis of variance test. But sometimes the analyst may be interested in a particular contrast or set of contrasts *before* the analysis of variance is carried out, or instead of the analysis of variance altogether. For example, with the swimming data, we may only be interested in whether swimming frequencies among rural residents differ from all the others. With the precipitation data, we may only wish to know whether weekend and weekday magnitudes differ.

When contrasts are specified prior to the analysis of variance, confidence intervals are narrower than when they are determined using Equation 4.23, after the fact. For a given contrast, the *a priori* confidence interval at the $1-\alpha$ level is

$$\psi - t_{N-k}\,\sigma_\psi \le \psi \le \psi + t_{N-k}\,\sigma_\psi \tag{4.26}$$

where σ_ψ is defined as before, and the critical value of t comes from a t-table with $N-k$ degrees of freedom, using $\alpha/2$ in each tail. If we are interested in more than one contrast, the value of α has to be apportioned among the contrasts of interest. For example, if we were interested in looking at five *a priori* contrasts, we could use $\alpha = 0.01$ for each of the five contrasts, giving a simultaneous confidence level of 0.95.

An example illustrating the use of contrasts is given in Section 4.8.

4.7 Spatial Dependence

One of the assumptions in ANOVA is that the observations within each category are independent. With spatial data, observations are often dependent, and some adjustment to the analysis should be made. The general effect of spatial dependence will be to render the effective number of observations smaller than the actual number of observations. With an effectively smaller number of observations, results are not as significant as they appear in the F-tests outlined in this chapter. With spatial data, therefore, it is possible that significant findings are due to the spatial dependence among the observations, and not to any real underlying differences in the means of the categories.

Griffith (1978) has proposed a spatially adjusted ANOVA model. The details of his model are beyond the scope of this text. Griffith's paper may also be of interest since it contains citations to other studies in geography that use analysis of variance.

4.8 One-Way ANOVA in *SPSS for Windows 9.0*

4.8.1 Data Entry

As before, there is a separate row for each observation. Using the data in Table 4.2, we will have 24 rows and 2 columns. Again, the second column designates the group number, and now we have added a third value to correspond with the rural region. The data are entered into the Data Editor of *SPSS* as follows:

```
38 1
42 1
50 1
57 1
80 1
70 1
32 1
20 1
58 2
66 2
80 2
62 2
73 2
39 2
73 2
58 2
80 3
70 3
60 3
55 3
72 3
73 3
81 3
50 3
```

4.8.2 Data Analysis and Interpretation

The analysis of variance proceeds in *SPSS 9.0 for Windows* by clicking on Analyze, then on Compare Means, and then on One-Way ANOVA. Swim (or whatever name is given to the variable in the first column) is then highlighted and moved over into the dialog box entitled Dependent List, and Location is highlighted and moved over into the dialog box entitled Factor. At this point, one can simply click OK to proceed with the analysis, but here we will also click on Options, and then check the boxes entitled Descriptive and Homogeneity of Variance. Also, *post hoc* contrasts can be made by simply clicking on Post Hoc and then clicking on the box labeled Scheffé. *A priori* contrasts are chosen by clicking on the box labeled Contrasts. Suppose we wish to contrast swimming frequency in the central city with the average swimming frequency in the other two regions. After choosing Contrasts, click on Polynomial, and leave Linear as the selected polynomial. We then need to specify the contrast coefficients (the c's in Equation 4.20). Here we could use either $c_1 = 1, c_2 = -0.5$, and $c_3 = -0.5$, or $c_1 = -1, c_2 = 0.5$, and $c_3 = 0.5$. Enter the coefficients one at a time, clicking on Add after each entry. Finally, choose Continue, and then OK.

The output that results is shown below.

Descriptives

SWIMFREQ

	N	Mean	Std. Deviation	Std. Error	95% Confidence Interval for Mean Lower Bound	Upper Bound	Minimum	Maximum
1.00	8	48.6250	19.8778	7.0278	32.0068	65.2432	20.00	80.00
2.00	8	63.6250	12.6597	4.4759	53.0412	74.2088	39.00	80.00
3.00	8	67.6250	11.4260	4.0397	58.0726	77.1774	50.00	81.00
Total	24	59.9583	16.6902	3.4069	52.9107	67.0060	20.00	81.00

Test of Homogeneity of Variances

SWIMFREQ

Levene Statistic	df1	df2	Sig.
1.509	2	21	.244

ANOVA

SWIMFREQ

	Sum of Squares	df	Mean Square	F	Sig.
Between Groups	1605.333	2	802.667	3.510	.048
Within Groups	4801.625	21	228.649		
Total	6406.958	23			

Contrast Tests

		Contrast	Value of Contrast	Std. Error	t	df	Sig. (2-tailed)
VAR00001	Assume equal variances	1	17.0000	6.5476	2.596	21	.017
	Does not assume equal	1	17.0000	7.6471	2.223	9.648	.051

Post Hoc Tests

Multiple Comparisons

Dependent Variable: VAR00001
Scheffe

(I) VAR00002	(J) VAR00002	Mean Difference (I-J)	Std. Error	Sig.	95% Confidence Interval	
					Lower Bound	Upper Bound
1.00	2.00	-15.0000	7.5606	.165	-34.9083	4.9083
	3.00	-19.0000	7.5606	.063	-38.9083	.9083
2.00	1.00	15.0000	7.5606	.165	-4.9083	34.9083
	3.00	-4.0000	7.5606	.870	-23.9083	15.9083
3.00	1.00	19.0000	7.5606	.063	-.9083	38.9083
	2.00	4.0000	7.5606	.870	-15.9083	23.9083

The first box again provides descriptive information on the variable in each region. Note that the mean frequency among respondents in the rural region is higher (67.625) than that in other regions, and its standard deviation is lower (11.426).

The second box gives us the results of a test of the assumption of homoscedasticity. This Levene's test supports the null hypothesis that the variances of the three region's responses could be equal (since the column headed "Sig." has an entry greater than 0.05) and that we have merely observed sampling variation. Had the *p*-value associated with this test been less than 0.05, we would have had to take the results of the analysis of variance more cautiously, since one of the underlying assumptions would have been violated.

The next box displays the results of the analysis of variance. The table gives the sums of squares, the mean squares, the degrees of freedom, and the *F*-statistic. Note that these match the results discussed in section 4.2.1, with small differences due to rounding error (as they should!). Importantly, the output also includes the *p*-value associated with the test under the column labelled "Sig.". Since this value is less than 0.05, we reject the null hypothesis, and conclude that there are significant differences in swimming frequencies among the residents of these three regions and that these differences cannot be attributed to sampling variation alone (unless we just happened to get a fairly unusual sample).

The results of the *a priori* (not shown) contrasts indicate that there is indeed a significant difference between the swimming frequencies in the central city and in other areas. The value of the contrast is 17, which is the mean difference in swimming frequencies $(65.63 - 48.63)$. The significance or *p*-value is indicated in the last column, and this is less than 0.05 when variances are assumed equal (and equal to .051 when variances are not assumed equal). Results of the *post hoc* contrasts indicate that there is one paired difference that is close to being significant – that between central city and rural regions. This is indicated by the p-value (in the column headed "Sig.") of .063. Confidence intervals for the difference in swimming frequencies associated with each paired comparison are also given.

4.8.3 Levene's Test for Equality of Variances

Levene's test of the assumption that the variances of each column of data are equal is actually similar to an analysis of variance test, except that the test is carried out on the absolute value of the data after the column means have been subtracted

Let

$$z_{ij} = |x_{ij} - x_{+j}|$$

Then Levene's statistic is

$$L = \frac{\sum_{j=1}^{k} n_j (z_{+j} - z_{++})/(k-1)}{\sum_{j=1}^{n_j} \sum_{i=1}^{k} (z_{ij} - z_{+j})^2 / \sum_{j=1}^{k} (n_j - 1)}$$

where there are k categories, z_{++} is the overall mean of the z's, and z_{+j} is the mean of the z's in category j. When the null hypothesis of equal column variances is true, Levene's statistic has an F-distribution, with $(k-1)$ and $\sum_{i=1}^{k}(n_i - 1)$ degrees of freedom.

Illustration. For the data in Table 4.2 on swimming frequencies, the first step is to subtract the column mean from each observation. Then take the absolute values of the results; these are the z-values. Then the required quantities are

$$k = 3; \qquad \sum_{j=1}^{3} (n_j - 1) = 21$$

$$z_{++} = 11.489; \qquad z_{+1} = 15.625; \qquad z_{+2} = 9.375; \qquad z_{+3} = 9.4675$$

$$\sum_{i=1}^{n_1} (z_{i1} - z_{+1})^2 = 812.75; \qquad \sum_{i=1}^{n_2} (z_{i2} - z_{+2})^2 = 418.75;$$

$$\sum_{i=1}^{n_3} (z_{i3} - z_{+3})^2 = 196.81$$

$$L = \frac{\left[8(15.625 - 11.489)^2 + 8(9.375 - 11.489)^2 + 8(9.4675 - 11.489)^2\right] \Big/ (3-1)}{(812.75 + 418.75 + 196.81)/(21)}$$

$$= 1.509$$

This value of L is the same as that shown in the output, and it is not significant since it is less than the critical value of $F_{2,21} = 3.47$. Hence the assumption of homoscedasticity is satisfied.

4.8.4 Tests of Normality: the Shapiro–Wilk Test

One of the assumptions that we should test is whether the data come from a normal distribution. The Shapiro–Wilk test is a particularly good test of normality to use when sample sizes are small. Within *SPSS for Windows 9.0*, it can be run by using Analyze/Descriptive Statistics/Explore. Before clicking on OK, click on Plots, and check the box that reads "Normality plots with tests".

The results will include a p-value for the Shapiro–Wilk (W) statistic. The W statistic is found as

$$W = \frac{b^2}{\sum_{i=1}^{n} (x_i - \bar{x})^2} = \frac{b^2}{(n-1)s^2}$$

where

$$b = \sum_{i=1}^{k} a_{n-i+1}\{x_{(n-i+1)} - x_{(i)}\}$$

Here $k = n/2$ when n is even, and the x-values have been ordered so that $x_{(1)} < x_{(2)} < \cdots < x_{(n)}$. The α coefficients come from a table, and one is provided in Table A.6 in Appendix A. If W is less than its critical value (also taken from a table; see Table A.7), the null hypothesis of normality is rejected. For further details, see Shapiro and Wilk (1965).

Illustration. To determine whether the eight swimming frequencies observed in the rural area could have come from a normal distribution, we use

$$x_{(1)} = 50, \quad x_{(2)} = 55, \quad x_{(3)} = 60, \quad x_{(4)} = 70,$$
$$x_{(5)} = 72, \quad x_{(6)} = 73, \quad x_{(7)} = 80, \quad x_{(8)} = 81$$

From the table, for $n = 8$ and $k = 4$, $\alpha_8 = .6052$, $\alpha_7 = .3164$, $\alpha_6 = .1743$, and $\alpha_5 = .0561$. Thus

$$b = .6052(81 - 50) + .3164(80 - 55) + .1743(73 - 60) + .0561(72 - 70) = 29.04$$

and

$$W = \frac{29.04^2}{(n-1)s^2} = \frac{843.86}{913.875} = .923$$

Since this is greater than the critical value of $W_{.05} = .818$, we accept the null hypothesis and conclude that there is not enough evidence to reject the assumption of normality.

Exercises

1. Using the following data:

Precipitation at Boston Airport (inches)

Year	SAT	SUN	MON	TUE	WED	THUR	FRI
1971 II	0.83	3.14	4.20	1.28	1.16	4.25	2.08
1972 I	4.66	4.15	3.40	1.74	3.91	5.15	5.06
1972 II	3.03	5.80	2.29	3.17	3.50	3.40	3.04
1973 I	3.69	3.72	4.29	2.06	3.04	2.30	4.26
1973 II	2.35	3.62	3.56	2.27	4.46	2.52	3.36
1974 I	3.18	3.28	1.82	3.75	2.07	3.54	2.27

(a) Find the mean and standard deviation for each day of the week.
(b) Use Levene's test to determine whether the assumption of homoscedasticity is justified.
(c) Perform an analysis of variance to test the null hypothesis that precipitation does not vary by day of the week. Show the between and within sum of squares, the observed F-statistic, and the critical F-value.
(d) Repeat the analysis using data for Pittsburgh:

Precipitation at Pittsburgh Airport (inches)

Year	SAT	SUN	MON	TUE	WED	THUR	FRI
1971 II	1.64	5.55	3.19	2.45	1.44	1.07	1.66
1972 I	2.20	3.37	0.78	2.63	2.32	5.57	2.80
1972 II	2.75	1.72	2.34	3.40	3.68	3.48	2.50
1973 I	2.23	4.31	2.02	1.83	4.35	4.07	2.66
1973 II	3.65	2.66	3.95	2.31	1.85	2.63	1.11
1974 I	4.96	3.00	2.61	1.75	2.70	2.45	4.06

2. Assume that an analysis of variance is conducted for a study where there are $N = 50$ observations and $k = 5$ categories. Fill in the blanks in the following ANOVA table:

	Sums of squares	Degrees of freedom	Mean square	F
Between	—	—	116.3	—
Within	2000	—	—	
Total	—	—		

If the critical value of F is 2.42, what is your conclusion regarding the null hypothesis that the means of the categories are equal?

3. What are the assumptions of analysis of variance? What does it mean to say that analysis of variance is relatively robust with respect to deviations from the assumptions? What does it mean to say that the Kruskal–Wallis test is not as powerful as ANOVA?

4. Fill in the blanks in the following analysis of variance table. Then compare the F value with the critical value, using $\alpha = 0.05$.

	Sums of squares	df	Mean square	F
Between SS	34.23	2	—	—
Within SS	—	—	—	
Total SS	217.34	35		

5. Using

$$\text{WSS} = \sum_{i=1}^{k} (n_i - 1)\, s_i^2$$

(a) Find the within sum of squares for the following data:

Toxin levels in shellfish (mg)

Observation	Long Island Sound	Great South Bay	Shinnecock Bay
1	32	54	15
2	23	27	18
3	14	18	19
4	42	11	21
5	13	10	28
6	22	34	9
Mean	24.33	25.67	18.33
Std. dev.	11.08	16.69	6.31
Overall mean = 22.78. Overall std. dev. = 11.85			

(b) Find the value of the test statistic F and compare it with the critical value.

(c) Rank the data (1 = lowest), using the average of the ranks for any set of tied observations. Then find the Kruskal–Wallis statistic

$$H = \left(\frac{12}{N(N+1)} \sum_{i=1}^{k} \frac{R_i^2}{n_i} \right) - 3(N+1)$$

Then adjust the value of H by dividing it by

$$1 - \frac{\sum_i (t_i^3 - t_i)}{N^3 - N}$$

where t_i is the number of observations that are tied for a given set of ranks. Compare this test statistic with the critical value of chi-square, which has $k - 1$ degrees of freedom to decide whether to accept or reject the null hypothesis.

6. Twelve low-income and twelve high-income individuals are asked about the distance of their last residential move. The following data represent the distances moved, in kilometers:

	Low income	High income
	5	25
	7	24
	9	8
	11	2
	13	11
	8	10
	10	10
	34	66
	17	113
	50	1
	17	3
	25	5
Mean	17.17	23.17
Std. dev.	13.25	33.45

Test the null hypothesis of homogeneity of variances by forming the ratio s_1^2/s_2^2, which has an F-ratio with $n_1 - 1$ and $n_2 - 1$ degrees of freedom. Then use the appropriate F-test. Set up the null and alternative hypotheses, choose a value of alpha and a test statistic, and test the null hypothesis. What assumption of the test is likely not satisfied?

7. Are the confidence intervals associated with *a priori* contrasts in ANOVA narrower or wider than *a posteriori* contrasts? Why? Which would be more powerful in rejecting the null hypothesis that the contrast was equal to zero?

8. A study groups 72 observations into nine groups, with eight observations in each group. The study finds that the variance among the 72 observations is 803. Complete the following ANOVA table:

	Sums of squares	df	Mean square	F
Between	6000	—	—	—
Within	—	—	—	
Total	—			

If the critical value of F is 2.8, what do you conclude about the hypothesis that the means of all groups are equal? What can you conclude about the p-value?

9. A sample is taken of incomes in three neighborhoods, yielding the following data:

	Neighborhood			Overall
	A	B	C	(combined sample)
n	12	10	8	30
mean	43.2	34.3	27.2	35.97
std. dev.	36.2	20.3	21.4	29.2

Use analysis of variance to test the null hypothesis that the means are equal.

10. Use the Kruskal–Wallis test to determine whether you should accept the null hypothesis that the means of the following four columns of data are equal:

Col 1	Col 2	Col 3	Col 4
23.1	43.1	56.5	10002.3
13.3	10.2	32.1	54.4
15.6	16.2	43.3	8.7
1.2	0.2	24.4	54.4

11. A researcher is interested in differences in travel behavior for residents living in four different regions. From a sample of size 48 (12 in each region), she finds that the mean commuting distance is 5.2 miles, and that the standard deviation is 3.2 miles. What is the total sum of squares? Suppose that the standard deviations for each of the four regions are 2.8, 2.9, 3.3, and 3.4. What is the within sum of squares? Fill in the table:

	Sum of squares	df	Mean square	F
Between	—	—	—	—
Within	—	—	—	
Total	—			

Suppose the critical value of F is 2.7. Do you accept or reject the null hypothesis?

12. A researcher wishes to know whether distance travelled to work varies by income. Eleven individuals in each of three income groups are surveyed. The resulting data are as follows (in commuting miles, one-way):

	Income		
Observations	Low	Medium	High
1	5	10	8
2	4	10	11
3	1	8	15
4	2	6	19
5	3	5	21
6	10	3	7
7	6	16	7
8	6	20	4
9	4	7	3
10	12	3	17
11	11	2	18

Use analysis of variance to test the hypothesis that commuting distances do not vary by income. Also evaluate (using, e.g., the Levene test) the assumption of homoscedasticity. Finally, lump all of the data together and produce a histogram, and comment on whether the assumption of normality appears to be satisfied.

13. Data are collected on automobile ownership by surveying residents in central cities, suburbs, and rural areas. The results are:

	Central cities	Suburbs	Rural areas
No. of observations	10	15	15
Mean	1.5	2.6	1.2
Std. dev.	1.0	1.1	1.2
Overall mean = 1.725. Overall std. dev. = 1.2.			

Test the null hypothesis that the means are equal in all three areas.

5 Correlation

LEARNING OBJECTIVES

- Understanding the nature of the relationship between two variables
- Understanding the effects of sample size on tests of significance
- Alternative tests of correlation when assumptions are not reasonable

5.1 Introduction and Examples of Correlation

One of the most common objectives of researchers is to determine whether two variables are associated with one another. Does patronage of a public facility vary with income? Does interaction vary with distance? Do housing prices vary with accessibility to major highways? Researchers are interested in how variables *co-vary*.

The concept of *covariance* is a straightforward extension of the concept of variance. Whereas the variance is the expected or average value of the squared deviation of observations on a single variable from their mean, the covariance is the expected or average value of the product of the two variables' (say X and Y) respective deviations from their means:

$$\text{Cov}[X, Y] = E[(X - \mu_X)(Y - \mu_Y)] \tag{5.1}$$

The reader may note that, using an argument outlined in Appendix B, Equation 5.1 may be rewritten

$$E[(X - \mu_X)(Y - \mu_Y)] = E[XY] - \mu_X \mu_Y \tag{5.2}$$

Alternatively, based upon Equation 5.2, the sample covariance may be found using

$$\text{Cov}[X, Y] = \frac{1}{n}\sum_{i=1}^{n} x_i y_i - \bar{x}\bar{y} \tag{5.3}$$

Note that the covariance of a variable X with itself, Cov(X, X), is equal to the variance of x. In practice, the covariance may be found by taking the average of the products of the deviations from the means (although using $n-1$ instead of n in the denominator, as is the case with the variance):

$$\text{Cov}(X, Y) = \frac{1}{n-1}\sum_{i=1}^{n}(x_i - \bar{x})(y_i - \bar{y}) \tag{5.4}$$

The covariance of X and Y may be negative or positive. The covariance will be positive if most of the points (x, y) lie along a line with positive slope when they are plotted. The covariance will be negative when the plotted points lie along a line with negative slope. Figure 5.1 depicts points in the (x, y) plane, where the axes are centered at $(\bar{x}, \bar{y})$. It demonstrates that points lying in quadrants I and III will contribute positively to the covariance, and points in quadrants II and IV will contribute negatively to the covariance.

The magnitude of the covariance will depend upon the units of measurement. The covariance may be standardized, so that its values lie in the range from -1 to $+1$, by dividing by the product of the standard deviations. This standardized covariance is known as Pearson's *correlation coefficient*. The correlation coefficient provides a standardized measure of the linear association between two variables. Its theoretical value is

$$\rho = \frac{\Sigma[(X - \mu_X)(Y - \mu_Y)]}{\sigma_X \sigma_Y} \tag{5.5}$$

where σ_x and σ_y refer to the standard deviation of variables x and y in the population. The sample correlation coefficient, r, may be found from

$$r = \frac{\sum_{i=1}^{n}(x_i - \bar{x})(y_i - \bar{y})}{(n-1)s_x s_y} \tag{5.6}$$

where s_x and s_y are the sample standard deviations of variables x and y, respectively. This is known as *Pearson's* correlation coefficient. Note that this is equal to

$$r = \frac{\sum_{i=1}^{n} z_x z_y}{n-1} \tag{5.7}$$

where z_x and z_y are the z-scores associated with x and y, respectively.

It is important to note that the correlation coefficient is a measure of the strength of the *linear* association between variables. As Figure 5.2 demonstrates, it is possible to have a strong, nonlinear association between two variables and yet have a correlation coefficient close to zero. One implication of this is that it is important to plot data (the term *scatterplot* is often used to refer to graphs such as Figures 5.1 and 5.2, where each observation is represented by a point in the plane, and where the two axes represent the levels of the two variables), since potential associations between the variables might be revealed in those cases where the value of r is low.

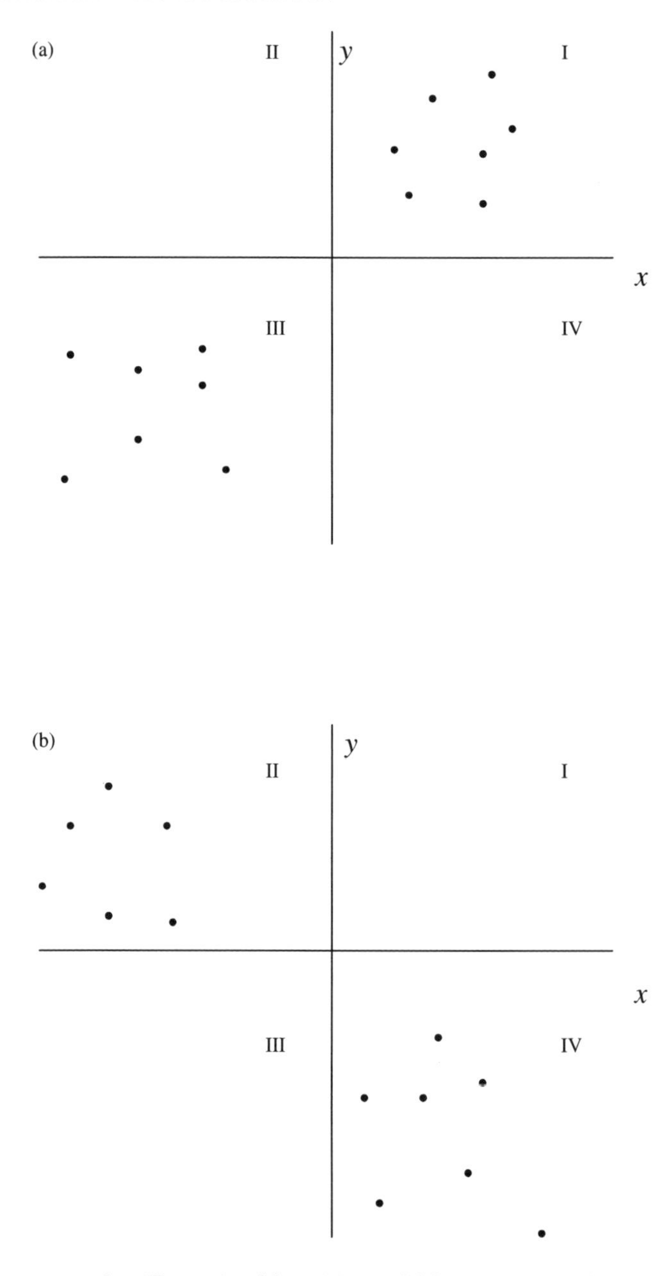

Figure 5.1 **Scatterplots illustrating (a) positive and (b) negative correlation**

It is also important to realize that the existence of a strong linear association does not necessarily imply that there is a *causal* connection between the two variables. A strong correlation was once found between British coal production and the death rate of penguins in the Antarctic, but it would be a stretch of the imagination to connect the two in any direct way! Changes in both British coal production and the death rate of penguins happened to go in the same

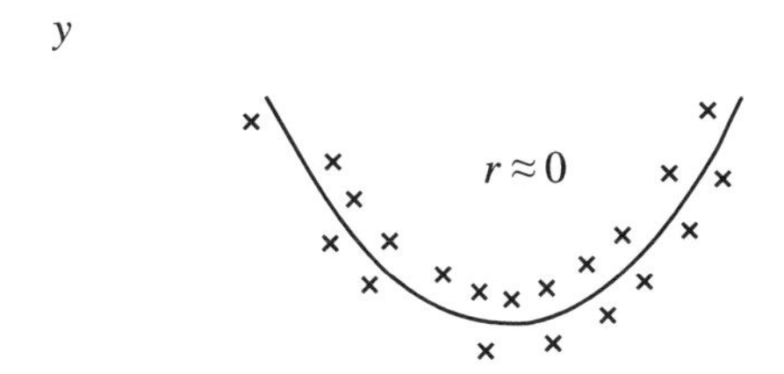

Figure 5.2 **Nonlinear relationship with *r* approximately 0**

direction over a period of time, but this does not necessarily imply a causal connection between the two. Another article once pointed out the strong connection between the annual number of tornados and the volume of automobile traffic in the United States. The claim – presumably in jest – was that both the number of tornados and the volume of automobile traffic had steadily increased in the United States throughout the twentieth century. If years were used as observations on the x-axis, with the number of tornados on the y-axis, a very strong positive correlation would be observed; years with many tornados would coincide with years with a high volume of traffic. Strong linear relationships often prompt deep thought about possible explanations, and in this case an explanation was offered. The correlation was deemed to be due to the fact that Americans drive on the right-hand side of the road! As cars pass one another, counterclockwise movements of air are generated, and we all know that counterclockwise movements of air are associated with low pressure systems. Some of these low pressure systems spawn tornados. Increasing traffic, then, would understandably lead to more tornados. Furthermore, since the British drive on the left, it should come as no surprise that there are not many tornados there! (Though I doubt one could claim that they have the great weather one would expect from the high pressure systems created by the clockwise movement of traffic-generated air currents!). A better explanation of the relationship is that the two have increased over time for very different reasons. The increase in the number of tornados is likely due to the simple fact that the weather observation network is better than it used to be. The search for a causal relationship is an important one, but the effort may sometimes be carried too far!

5.2 More Illustrations

5.2.1 Mobility and Cohort Size

Easterlin (1980) has suggested that young adults who are members of a large cohort (like the baby boom) will face a more difficult time in labor and housing markets. For these cohorts, the supply of people relative to the number of job and housing opportunities is relatively high. Consequently, there will be a tendency for mortgage rates and unemployment to be higher when large cohorts pass through their young adult years. Similarly, mortgage and unemployment rates will tend to be lower when small cohorts reach their twenties and thirties. Rogerson (1987) has extended this argument to hypothesize that large cohorts of young adults will exhibit lower mobility rates, since the cohort's opportunities for changing residence will be limited by the relatively inferior state of the labor and housing markets. The mobility rate is measured as the percentage of individuals changing residence during a one-year period, and the size of the young adult cohort is measured by the fraction of the total population in a specified young adult age group. Data on these variables for the period 1948–84 are presented in Table 5.1.

For 20–24 year olds, $n = 28$, and the correlation coefficient between mobility rate and cohort size is equal to -0.747; for 25–29 year olds, $r = -0.805$. For the 20–24 year olds, the cross-product $\sum(x_i - \bar{x})(y_i - \bar{y})$ (i.e., the numerator

Table 5.1 **U.S. mobility data, 1948–1984**

	Mobility rate		Fraction of total population	
Year	20–24	25–29	20–24	25–29
1948–49	35.0	*	.0804	.0829
1949–50	34.0	*	.0784	.0821
1950–51	37.7	33.6	.0767	.0812
1951–52	37.8	31.6	.0746	.0794
1952–53	40.5	33.4	.0720	.0774
1953–54	38.1	30.5	.0757	.0762
1954–55	41.8	31.3	.0735	.0758
1955–56	44.5	32.3	.0713	.0750
1956–57	41.2	32.0	.0694	.0734
1957–58	42.6	34.6	.0671	.0715
1958–59	42.5	33.2	.0645	.0701
1959–60	41.2	32.1	.0617	.0682
1960–61	43.6	34.4	.0616	.0605
1961–62	43.2	33.0	.0625	.0592
1962–63	42.0	34.6	.0641	.0582
1963–64	43.4	35.2	.0672	.0580
1964–65	45.0	35.8	.0692	.0528
1965–66	42.4	35.5	.0708	.0584
1966–67	41.0	33.0	.0715	.0593
1967–68	41.5	33.2	.0767	.0609
1968–69	42.5	32.6	.0787	.0638
1969–70	41.8	32.6	.0813	.0658
1970–71	41.2	32.4	.0839	.0669
1975–76	38.0	32.6	.0897	.0790
1980–81	36.8	30.1	.0943	.0859
1981–82	35.5	30.0	.0949	.0868
1982–83	33.7	29.8	.0939	.0892
1983–84	34.1	30.1	.0928	.0904

Note: Data unavailable for missing years.
Source: Rogerson (1987).

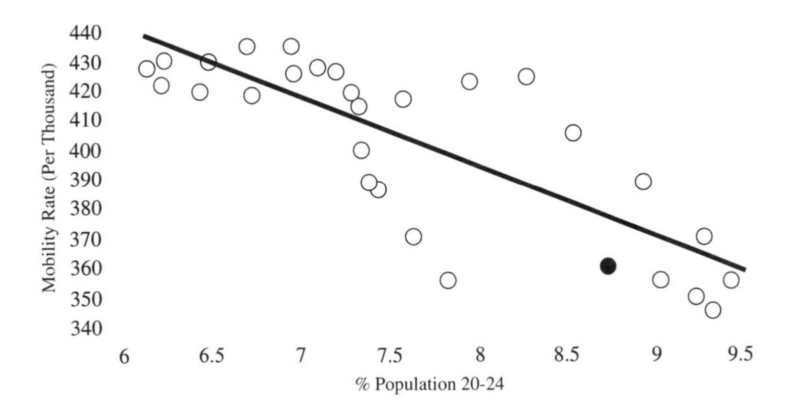

Figure 5.3 **Correlation of mobility with cohort size. Source: Plane and Rogerson (1991)**

of the covariance) is $-.694$. Dividing by $n-1=27$ yields a covariance of $-.0257$. The standard deviations for x and y are 3.371 and 0.0102, respectively. Dividing the covariance by the product of these standard deviations, as in Equation 5.6, yields the correlation coefficient of $-.747$. The data for 20–24 year olds are graphed in Figure 5.3, where the negative relationship between the variables is apparent.

5.2.2 Statewide Infant Mortality Rates and Income

As part of an assignment back in my graduate school days, I decided to investigate geographic variation in infant mortality rates in the United States. The data I collected were at the state level. I was interested in understanding whether infant mortality rates varied with factors such as educational attainment, income, access to health care, etc. As part of my analysis, I graphed the relationship between infant mortality rates and personal income for the white population. The graph is shown in Figure 5.4. Most of the states fall close to a line with negative slope, ranging from states such as Mississippi and Kentucky (with low values of personal income and high infant mortality rates) to states such as Connecticut, where the statewide personal income was high and mortality rates were low. Pearson's correlation coefficient for the 50 states is equal to -0.28. Notice, though, the presence of six states that have infant mortality levels above the level expected, given the personal income in the state (TX, CO, AZ, WY, NM, and NV). Cases such as this that do not fit the general trend are known as *outliers*. It is interesting to note that these states form a compact geographic cluster. Additional inspection of the data had raised the possibility that these six states had a relatively low number of physicians per 100 000 statewide residents. A two-tail t-test confirmed that indeed these six states had a significantly lower number of physicians per 100 000, relative to the other states.

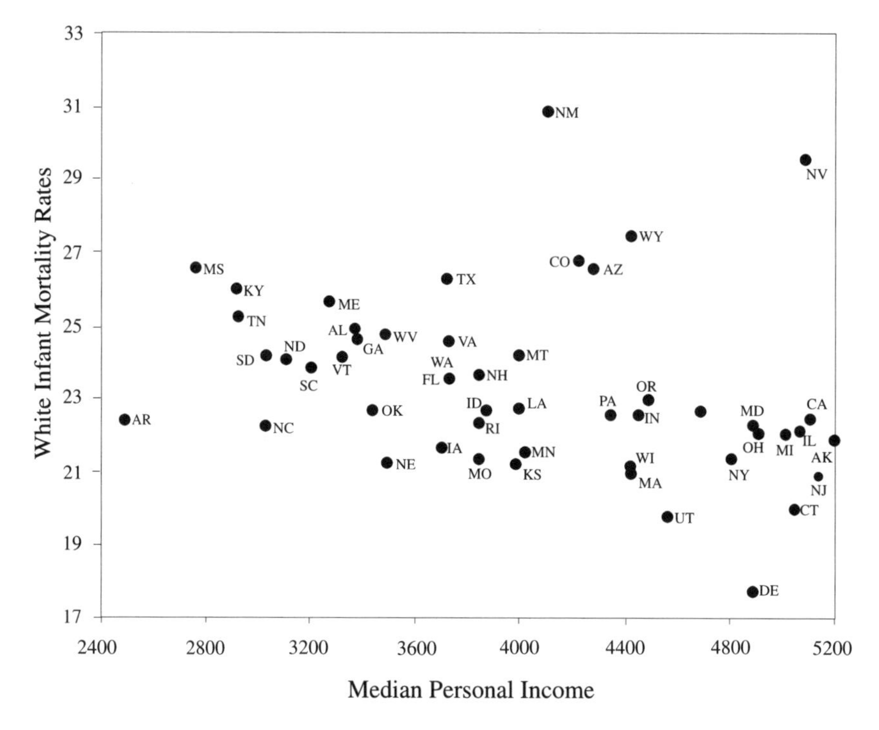

Figure 5.4 **White infant mortality rates as a function of median personal income**

The treatment of outliers depends upon the circumstances. A good underlying understanding of *why* particular points are outliers provides some rationale for removing those points from the analysis. In this case, we have a reasonably good explanation for the outliers, and we are justified in asking what the correlation would be without the outliers (it is $r = -0.64$, a much stronger negative relationship than was found with the original 50 observations). Of course we would not want to get in the habit of plotting variables and arbitrarily eliminating those points that do not fall close to the line, just so that we can report a high value of r. But it *is* good practice to plot the data and think carefully about the reasons for any outliers.

5.3 A Significance Test for *r*

To test the null hypothesis that the true correlation coefficient, ρ, is equal to zero, the data for each variable are assumed to come from normal distributions. If this assumption is satisfied, the test may be carried out by forming the t-statistic

$$t = \frac{r\sqrt{n-2}}{\sqrt{1-r^2}} \tag{5.8}$$

If the null hypothesis is true, this statistic has a t-distribution, with $n-2$ degrees of freedom.

5.3.1 Illustration

The data in Table 5.1 for the $n=28$ observations for 20–24 year olds yields a correlation coefficient of $r=-0.747$. For the null hypothesis $H_0 : \rho = 0$, and a two-tailed test with $\alpha=0.05$, the t-statistic is

$$t = \frac{(-0.747)\sqrt{28-2}}{\sqrt{1-(-0.747)^2}} = -5.73 \tag{5.9}$$

A t-table reveals that the critical values of t, using $\alpha=0.05$ in a two-tail test with 26 degrees of freedom, are ± 2.056. The null hypothesis that the correlation coefficient is zero is rejected.

5.4 The Correlation Coefficient and Sample Size

An extremely important point is that the correlation coefficient is influenced by sample size. It is far easier to reject the null hypothesis that $\rho=0$ with a large sample size than it is with a small sample size. To see this, compare the situation where $r=0.4$ with a sample size of 11, and the situation where $r=0.4$ with a sample size of $n=38$. In the former case, the observed t-statistic is 1.3, which is less than the critical value $t_{0.05,9}=2.262$, and the null hypothesis is accepted. In the latter case, when $n=38$, the t-statistic is 2.0, and the null hypothesis is rejected since the t-statistic is greater than the critical value, $t_{0.05,36}=1.96$.

One of the implications of this is that there really should be no popular rules of thumb that are invoked to decide whether r is sufficiently high to make the researcher happy about the level of correlation. Such rules of thumb do seem to exist – for instance, an r value of 0.7 or 0.8 may be taken as important or significant. But, as we have just seen, whether a correlation coefficient is truly significant depends upon the sample size. Thus, when the researcher is working with large data sets, a relatively low value of r should not be as disappointing as that same value of r when the sample size is smaller. A value of $r=0.4$ could be quite meaningful if $n=1000$, and the researcher should not necessarily throw the results out the window just because the r-value is noticeably less than 1 and perhaps less than some arbitrary, rule-of-thumb value such as $r=0.8$. Table 5.2 gives, for various values of n, the minimum absolute value of r to achieve significance. For example, with a sample size of $n=50$, any value of $r>0.288$ or less than -0.288 would be found significant using the t-test described above. The reader will note how even quite small values of r are significant when the sample size is only modestly large. For values of $n>30$, the

Table 5.2 **Minimum values of r required for significance**

Sample size, n	Minimum absolute value of r needed to attain significance (using $\alpha=0.05$)
15	.514
20	.444
30	.361
50	.279
100	.197
250	.124

For large n, r_{crit} is approximately $2/\sqrt{n}$.

quantity $2/\sqrt{n}$ is equal to the approximate absolute value of r that is needed for significance using $\alpha=0.05$. For example, if $n=49$, a correlation coefficient with absolute value greater than $2/\sqrt{49}=0.286$ would be significant.

While we have just argued that we should not too hastily discard the results of an analysis because of a seemingly low correlation (since with a large sample size that correlation may be significantly different from zero), there is also some concern about attaching too *much* importance to the results of a significance test. Meehl (1990) has noted that with many data sets there is a strong tendency to find that "everything correlates to some extent with everything else" (p. 204). This is sometimes referred to as the "crud factor." There is no particular reason for believing that the correlation between any two variables chosen from most data sets should be *exactly* zero, and therefore, if the sample size is large enough, we will be able to reject the null hypothesis that they are unrelated. For example, Standing *et al.* (1991) find that in a data set containing 135 variables related to the educational and personal attributes of 2058 individuals, the typical variable exhibited a significant correlation with 41% of the other variables. The extreme case was the variable measuring Grade 5 mathematics scores – it was significantly correlated with 76% of the other variables, leading the authors to conclude that "the number of statistically significant possible 'causes' of mathematics achievement available to the unbridled theorizer will almost be as large" (p. 125).

5.5 Spearman's Rank Correlation Coefficient

In situations where only ranked data are available, or where the assumption of normality required for the test $H_0 : \rho=0$ is not satisfied, it is appropriate to use Spearman's rank correlation coefficient. As the name implies, this measure of correlation is based only upon the ranks of the data. Two separate sets of ranks are developed, one for each variable. A rank of 1 is assigned to the lowest value and a rank of n to the highest observation in each column. Spearman's rank correlation coefficient, r_s, is

$$r_S = 1 - \frac{6\sum_{i=1}^{n} d_i^2}{n^3 - n} \tag{5.10}$$

Table 5.3 **U.S. mobility data, 1948–84, with ranks**

	Mobility rate		Fraction of total population		
Year	20–24	Rank	20–24	Rank	d_i
1950–51	37.7	5	.0767	17.5	−12.5
1951–52	37.8	6	.0746	15	−9
1952–53	40.5	9	.0720	13	−4
1953–54	38.1	8	.0757	16	−8
1954–55	41.8	15.5	.0735	14	1.5
1955–56	44.5	25	.0713	11	14
1956–57	41.2	12	.0694	9	3
1957–58	42.6	21	.0671	6	15
1958–59	42.5	19.5	.0645	5	14.5
1959–60	41.2	12	.0617	2	10
1960–61	43.6	24	.0616	1	23
1961–62	43.2	22	.0625	3	19
1962–63	42.0	17	.0641	4	13
1963–64	43.4	23	.0672	7	16
1964–65	45.0	26	.0692	8	18
1965–66	42.4	18	.0708	10	8
1966–67	41.0	10	.0715	12	−2
1967–68	41.5	14	.0767	17.5	−3.5
1968–69	42.5	19.5	.0787	19	0.5
1969–70	41.8	15.5	.0813	20	−4.5
1970–71	41.2	12	.0839	21	−9
1975–76	38.0	7	.0897	22	−15
1980–81	36.8	4	.0943	25	−21
1981–82	35.5	3	.0949	26	−23
1982–83	33.7	1	.0939	24	−23
1983–84	34.1	2	.0928	23	−21

Source: Plane and Rogerson (1991).

where d_i^2 is the squared difference between the ranks for observation i, and n is the sample size. The mobility data for 20–24 year olds for the period 1950–1984 are repeated in Table 5.3, with ranks adjacent to the mobility and cohort size variables. Note that tied ranks are treated by replacing them with the average of the tied ranks. The differences in ranks, d_i, are given in the last column. For these data, $r_S = 1 - (6(5045.5))/(26^3 - 26) = -0.725$. To test hypotheses, we may use the fact that the quantity $r_S\sqrt{n-1}$ has a t-distribution with $n-1$ degrees of freedom. In our example, the observed value of t is therefore $-0.725\sqrt{26} = -3.70$. This is less than the critical value $t_{0.05,25} = -2.06$, and so the null hypothesis of no association is rejected. (Technically, when there are tied ranks, Equation 5.10 should not be used; instead one calculates Spearman's correlation coefficient by calculating Pearson's correlation coefficient, using the ranks as observations.)

5.6 Additional Topics

5.6.1 Confidence Intervals for Correlation Coefficients

We saw above that a t-statistic may be used to test the hypothesis that $\rho = 0$. One might suppose that the fact that this statistic has a t-distribution could be used to create a confidence interval for ρ in much the same way that confidence intervals are created for means or differences between means. Thus one might contemplate using the observed value of Pearson's r as follows:

$$r - t_{\alpha,n-2}\hat{\sigma}_r \leq \rho \leq r + t_{\alpha,n-2}\hat{\sigma}_r \tag{5.11}$$

where

$$\hat{\sigma}_r = \frac{\sqrt{1 - r^2}}{\sqrt{n - 2}} \tag{5.12}$$

The problem with this idea is that, in general, the sampling distribution of r is not symmetric, and therefore the confidence intervals will not be accurate.

A more accurate confidence interval may be constructed by first transforming the value of r into a variable that has a normal distribution. The quantity y is derived as follows:

$$y = 1.151 \ln \frac{1 + r}{1 - r} \tag{5.13}$$

This variable has a normal distribution with standard deviation equal to

$$\sigma_y = \frac{1}{\sqrt{n - 3}} \tag{5.14}$$

and so a 95% confidence interval for the quantity y is

$$y \pm y_{0.05}\,\hat{\sigma}_y = y \pm \frac{1.96}{\sqrt{n - 3}} \tag{5.15}$$

The endpoints of this confidence interval can then be transformed back into values of r using Equation 5.13. More specifically, Equation 5.13 can be solved for r:

$$r = \frac{e^{y/1.151} - 1}{e^{y/1.151} + 1} \tag{5.16}$$

To illustrate, let us place a confidence interval around the observed correlation of -0.747 between mobility rates and cohort size (where $n = 28$). We have

$$y = 1.151 \ln \frac{1 + (-0.747)}{1 - (-0.747)} = -2.224 \tag{5.17}$$

The lower and upper limits of y are therefore

$$-2.224 - \frac{1.96}{\sqrt{28-3}} = -2.616 \quad \text{and} \quad -2.224 + \frac{1.96}{\sqrt{28-3}} = -1.832 \qquad (5.18)$$

Equation 5.16 is then solved twice for the limits of the confidence interval for ρ, using each of these two values of y. We have

$$-0.813 \leq \rho \leq -0.662 \qquad (5.19)$$

Note that these limits are not symmetric around the observed value of -0.747.

5.6.2 Differences in Correlation Coefficients

The transformation of r into the quantity y is also used when comparing correlation coefficients with one another. Suppose we wish to know whether the correlation between two variables observed in one year is significantly different from the correlation observed in another year. The null hypothesis of no difference in the correlations, $H_0 : \rho_1 = \rho_2$, could be tested by converting the observed correlations r_1 and r_2 to y values, and then using the z-statistic

$$z = \frac{y_1 - y_2}{\sigma_{y_1 - y_2}} \qquad (5.20)$$

where

$$\sigma_{y_1 - y_2} = \sqrt{\frac{1}{n_1 - 3} + \frac{1}{n_2 - 3}} \qquad (5.21)$$

Is there a significant difference between the mobility/cohort size correlation coefficients for 20–24 year olds and 25–29 year olds? The transformed values, y, corresponding to the r-values of -0.747 and -0.805 are -2.224 and -2.561, respectively. The z-statistic is then

$$z = \frac{-2.224 - (-2.561)}{\sqrt{(1/(28-3)) + (1/(26-3))}} = 1.166 \qquad (5.22)$$

Comparing this with the critical value of $z_{0.05} = \pm 1.96$ used in a two-sided test implies that we accept the null hypothesis of no difference and conclude that the two correlation coefficients are not significantly different.

5.6.3 The Effect of Spatial Dependence on Significance Tests for Correlation Coefficients

The tests of significance outlined for both Pearson's r and Spearman's r_S assume that the observations of x are independent and that the values of y

are also independent. When the x and y variables come from spatial locations, this assumption of independence may not be satisfied. Indeed, one of the most important points in this book is that *spatial data often exhibit dependence* – the value of x in one location is often related to the value of x in nearby locations. In turn, spatial dependence affects the outcome of statistical tests, and this point should always be borne in mind when interpreting statistical results.

When spatial dependence is present and not accounted for, the variance of the correlation coefficient under the null hypothesis of no correlation is underestimated. When repeated samples are taken from spatially dependent data, and the x and y values follow the null hypothesis of no correlation between x and y, the frequency distribution of r values will look like the dotted line in Figure 5.5. The dotted line has a wider frequency distribution than the solid line. The solid line corresponds to the variability in r that is calculated as present when standard significance tests such as Equation 5.3 are applied. The critical values associated with the standard statistical tests (b and c) are lower in absolute value than those which *should* be used (a and d). When sample values of r fall in the shaded region, the standard statistical tests will incorrectly imply that the correlation coefficient is significantly different from zero. Correlation coefficients falling in the shaded region are likely *not* significant, and may be the result of the underlying spatial dependence exhibited by the x and y variables. Haining (1990a) states this as follows:

> The important issue here is not to use conventional procedures to test for the significance of the correlation coefficient, and to recognize that a large r (or r_S) value may be due to spatial correlation [i.e. dependence] effects ... The risks of inferring association between variables that is nothing other than the products of the spatial characteristics of the system are real and call for caution on the part of the user (p. 321).

To see the effects of spatial dependence on correlation tests, consider the following model for the value of a variable a, at location (x_1, y_1), applied to the

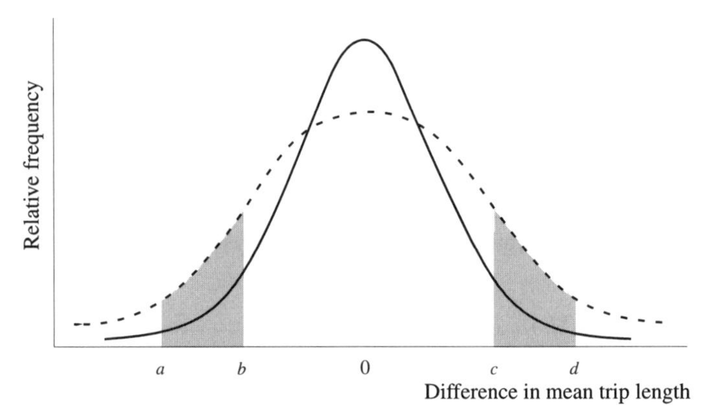

Figure 5.5 **Assumed (———) and actual (- - - -) variability in r when spatial dependence is present but not accounted for**

interior cells of a region that has been subdivided into a grid of square cells:

$$a(x_1, y_1) = \mu + 4\rho(\bar{a} - \mu) + \varepsilon(x_1, y_1) \quad (5.23)$$

where μ is the overall mean, $\bar{a}$ is the mean value of the variable in the four cells that share a side with (x_1, y_1), and $\varepsilon(x_1, y_1)$ is a normally distributed error term with mean zero. If $\rho = 0$, the values of the variable $a(x_1, y_1)$ are independent of the values at other sites. In this case the values of a are simply equal to the overall mean and a normally distributed error term. When $\rho > 0$, the value in a particular interior cell depends upon the values of the four surrounding cells. The value of ρ measures the amount of dependence, and it can range up to 0.25. Note that, when $\rho = 0.25$, the value of a at a location is precisely equal to the average of the values in surrounding locations, plus an error term. Clifford and Richardson (1985) use 5.23 to simulate two spatial variables that are not correlated with one another. Next, they find r and use Equation 5.8 and $\alpha = 0.05$ to see whether r is significant. One would expect to find significant values 5% of the time (since 0.05 is the Type I error probability). Table 5.4, as reported by Haining (1990a), displays the findings. ρ_1 and ρ_2 represent the amount of spatial dependence used in generating the two variables.

When ρ_1 and ρ_2 are zero, the Type I error probability is near its expected value of 0.05. Note that when one variable has no spatial dependence ($\rho_1 = 0$), the other variable can exhibit strong spatial dependence (e.g., $\rho_2 = 0.24$), and there is still no effect on the test for correlation since the Type I error probability is still near 0.05. But when both variables exhibit strong spatial dependence, the Type I error probabilities – where one incorrectly finds significant correlation coefficients – rise dramatically. With strong spatial dependence and no corrective action, one will too often reject true null hypotheses.

5.6.4 Modifiable Area Unit Problem and Spatial Aggregation

Gehlke and Biehl (1934) noted that correlation coefficients tend to increase with the level of geographic aggregation when census data are analyzed. A

Table 5.4 **Type I error probabilities with spatial dependence**

ρ_1	ρ_2	Type I error probability
0	0	0.0566
0	0.2	0.0500
0	0.24	0.0400
0.1	0.1	0.0700
0.1	0.225	0.1000
0.15	0.15	0.1000
0.15	0.225	0.1500
0.2	0.2	0.1900
0.2	0.24	0.3100
0.225	0.225	0.3366
0.24	0.24	0.5000

smaller number of large geographic units tends to give a larger correlation coefficient than does an analysis with a larger number of small geographic units. In a classic study, Robinson (1950) noted that the correlation between race and illiteracy rose with the level of geographic aggregation. It is important to keep in mind the fact that the size and configuration of spatial units may affect the analysis. What is significant at one spatial scale may not be significant at another.

5.7 Correlation in *SPSS for Windows 9.0*

Each variable should be represented by a column of data. Then click on Analyze and Correlate. Next, click on Bivariate, and move the variables you wish to correlate from the list on the left to the box on the right. You may move more than two variables into the box if you wish to see a table of correlations among a number of variables. If desired, check the box to have Spearman's correlation coefficient calculated (Pearson's correlation is calculated by default).

Table 5.5 **1990 Census data for a random sample of census tracts in Erie County, New York**

AREANAME		TOTPOP90	MEDHSINC	MEDAGE	SAGE	MAGE	PCTOWN
Tract 0010	BG 4	999	20862	49.24	19.08	60	.510
Tract 0016	BG 2	477	17804	50.70	17.49	60	.354
Tract 0026	BG 1	647	10545	51.24	16.92	58	.5
Tract 0028	BG 2	856	14602	50.66	18.29	60	.479
Tract 0045	BG 5	994	33603	45.12	15.36	60	.683
Tract 0057	BG 4	1083	24440	52.31	18.05	60	.516
Tract 0060	BG 1	879	15964	39.43	15.83	60	.416
Tract 0068	BG 2	374	33750	45.07	16.92	60	.202
Tract 0068	BG 4	806	14597	42.93	18.30	60	.346
Tract 0073.02	BG 9	2194	39779	52.49	16.58	39	.906
Tract 0076	BG 4	1150	29250	54.65	17.97	46	.884
Tract 0079.01	BG 3	1720	44205	53.63	13.83	41	.978
Tract 0079.02	BG 5	540	34625	60.55	14.06	44	.967
Tract 0079.02	BG 8	1128	32439	58.49	14.18	44	.899
Tract 0085	BG 1	434	39375	54.07	17.89	54	.967
Tract 0087	BG 3	1415	29513	51.42	17.77	60	.706
Tract 0097.01	BG 2	1639	39104	53.35	12.48	33	.992
Tract 0100.02	BG 1	3072	26174	54.41	16.36	30	.886
Tract 0100.02	BG 5	1715	28477	49.28	15.38	42	.638
Tract 0101.01	BG 5	755	35000	56.58	15.48	54	1
Tract 0101.02	BG 2	731	17647	43.49	16.66	45	.239
Tract 0111	BG 4	544	28438	57.78	17.04	47	.796
Tract 0115	BG 1	885	33214	51.68	19.54	47	.862
Tract 0117	BG 2	633	36346	48.22	16.09	47	.856
Tract 0120.02	BG 1	851	28500	59.56	17.39	42	.876
Tract 0142.05	BG 1	681	38125	46.44	18.23	19	.885
Tract 0150.03	BG 2	1270	25515	51.64	16.78	60	.58
Tract 0152.02	BG 9	1334	29554	49.15	17.89	37	.74
Tract 0153.02	BG 1	634	47083	52.38	15.21	46	.86

Table 5.6 **Bivariate correlations among four neighborhood variables**

Correlations

		MEDHSINC	SAGE	MEDAGE	PCTOWN
MEDHSINC	Pearson Correlation	1.000	−.428*	.362	.739**
	Sig. (2-tailed)	.	.021	.053	.000
	N	30	29	29	29
SAGE	Pearson Correlation	−.428*	1.000	−.242	−.370*
	Sig. (2-tailed)	.021	.	.207	.048
	N	29	29	29	29
MEDAGE	Pearson Correlation	.362	−.242	1.000	.684**
	Sig. (2-tailed)	.053	.207	.	.000
	N	29	29	29	29
PCTOWN	Pearson Correlation	.739**	−.370*	.684**	1.000
	Sig. (2-tailed)	.000	.048	.000	.
	N	29	29	29	29

* Correlation is significant at the 0.05 level (2-tailed).
** Correlation is significant at the 0.01 level (2-tailed).

Nonparametric Correlations

Correlations

			MEDHSINC	SAGE	MEDAGE	PCTOWN
Spearman's rho	MEDHSINC	Correlation Coefficient	1.000	−.433*	.347	.742**
		Sig. (2-tailed)	.	.019	.065	.000
		N	30	29	29	29
	SAGE	Correlation Coefficient	−.433*	1.000	−.261	−.403*
		Sig. (2-tailed)	.019	.	.172	.030
		N	29	29	29	29
	MEDAGE	Correlation Coefficient	.347	−.261	1.000	.745**
		Sig. (2-tailed)	.065	.172	.	.000
		N	29	29	29	29
	PCTOWN	Correlation Coefficient	.742**	−.403*	.745**	1.000
		Sig. (2-tailed)	.000	.030	.000	.
		N	29	29	29	29

*. Correlation is significant at the .05 level (2-tailed)
**. Correlation is significant at the 0.1 level (2-tailed).

5.7.1 Illustration

The data in Table 5.5 are a random sample of 29 census block groups (areas of about 1000 people) from Erie County, New York, in 1990. For each block group, there are data on population (totpop90), median household income (medhsinc), the median age of heads of households (medage), the standard deviation of the householder's age (sage), which is a measure of age mixing in the block group, the median age of housing (mage), and the percentage of housing that is owner-occupied (pctown). Rogerson and Plane (1998) discuss the age structure of householders in residential neighborhoods, and develop a model showing how age structure is related to variables such as age of the housing in the neighborhood, mobility, and homeownership.

Table 5.6 shows the bivariate correlation coefficients (Pearson and Spearman) among four of the variables. The median age of householders in block groups has a significant correlation with homeownership; as one might expect, the association is positive, and median age is higher in areas of higher homeownership. The variability of ages in a neighborhood (sage) is negatively related to income (high income neighborhoods are more homogeneous with respect to age) and negatively related to homeownership (where ownership is high, the ages of the householders are more homogeneous). Finally, note the similarity between Pearson's and Spearman's coefficients.

Exercises

1.
(a) Find the correlation coefficient, r, for the following sample data on income and education:

Observation	Income ($\$ \times 1000$)	Education (years)
1	30	12
2	28	12
3	52	18
4	40	16
5	35	16

(b) Test the null hypothesis $\rho = 0$.
(c) Find Spearman's rank correlation coefficient for these data.
(d) Test whether the observed value of r_S from part (c) is significantly different from zero.

2.
(a) Draw a graph depicting a situation where the correlation coefficient is close to zero, but there is a clear relationship between two variables.
(b) Draw a graph depicting a situation where there is a strong positive relationship between two variables, but where the presence of a small number of outliers makes the strength of the relationship less strong.

3. The t-statistic for testing the significance of a correlation coefficient is

$$t = \frac{r\sqrt{n-2}}{\sqrt{1-r^2}}$$

with $n - 2$ degrees of freedom. If the sample size is 36 and $\alpha = 0.05$, what is the smallest absolute value a correlation coefficient must have to be significant? What if the sample size is 80?

4. Find the correlation coefficient for the following data:

Obs.	X	Y
1	2	6
2	8	6
3	9	10
4	7	4

5.
(a) Why is a "rule of thumb" for the significance of a correlation coefficient (e.g., r^2 above 0.7 is significant) not a good idea?
(b) Why is a very large sample a "problem" in the interpretation of significance tests for the correlation coefficient?

6. Find the correlation coefficient between median annual income in the United States and the number of horse races won by the leading jockey, for the period 1984–1995:

Year	Median income	Number of races won by leading jockey
1984	35 165	399
1985	35 778	469
1986	37 027	429
1987	37 256	450
1988	37 512	474
1989	37 997	598
1990	37 343	364
1991	36 054	430
1992	35 593	433
1993	35 241	410
1994	35 486	317

Test the hypothesis that the true correlation coefficient is equal to zero. Interpret your results.

6 Introduction to Regression Analysis

LEARNING OBJECTIVES

- Modeling one variable as a linear function of another
- Fitting a straight line through a set of points plotted in two dimensions
- Assumptions of linear regression
- Relationship of regression to analysis of variance
- Testing the significance of the regression slope

6.1 Introduction

Whereas correlation is used to measure the strength of the linear association between variables, regression analysis refers to the more complete process of studying the causal relationship between a dependent variable and a set of independent, explanatory variables. Linear regression analysis begins by assuming that a linear relationship exists between the dependent variable (y) and the independent variables (x), proceeds by fitting a straight line to the set of observed data, and is then concerned with the interpretation and analysis of the effects of the x variables on y, and with the nature of the fit. An important outcome of regression analysis is an equation that allows us to predict values of y from values of x.

Regression analysis is used to specify and test a functional relationship between variables. As discussed in Chapter 1, the process of description often leads one to suspect that two or more variables are related. Once we specify *how* the variables are related, we have a model, which may be thought of as a simplification of reality. Regression analysis provides us with (a) a simplified view of the relationship between variables, (b) a way of fitting the model with our data, and (c) a means for evaluating the importance of the variables and the correctness of the model.

For example, we may wish to know whether the distance that adults live away from their parents is dependent upon education, whether snowfall is dependent upon elevation, whether infant mortality is related to income, or whether park attendance is related to the income of the population living within a certain distance of the park. In each case a good place to begin is to plot the data on a graph, and to measure the correlation between variables. Linear regression analysis takes this a step further by finding the best-fitting straight line through the set of points.

When there is just one independent, explanatory variable, as is the case in the examples above, we wish to fit a straight line through the set of data points; the equation of this line is

$$\hat{y} = a + bx \tag{6.1}$$

where $\hat{y}$ is the predicted value of the dependent variable, x is the observed value of the independent variable, a is the intercept (or point where the line intersects the vertical axis), and b is the slope of the line. The quantities a and b represent *parameters* describing the line, and these will be estimated from the data. This case, with one independent variable, is known as *simple regression* or *bivariate regression*, and it is depicted in Figure 6.1. We shall in this chapter confine our attention to this special case. More generally, *multiple regression* (treated in the next chapter) refers to the case where there is more than one independent variable.

The slope of the line, b, may be interpreted as the change in the dependent variable expected from a unit change in the independent variable. For example, suppose a regression of housing sale prices on the square footage of houses yielded the following equation:

$$\hat{p} = 30\,000 + 70s \tag{6.2}$$

where $\hat{p}$ is the predicted housing sales price, and s represents square footage. The slope in this equation is 70, and this means that an increase of one square foot leads, on average, to a \$70 increase in sales price. The price predicted for a

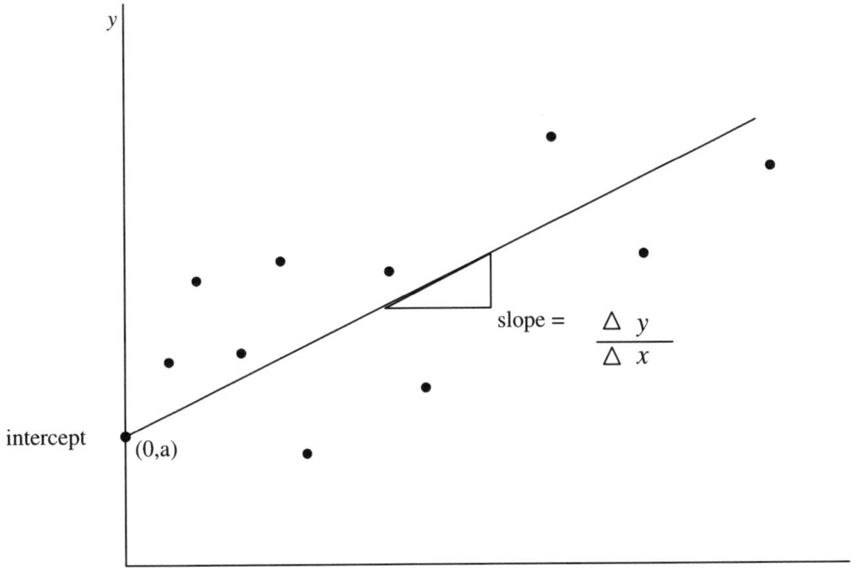

Figure 6.1 **Regression line through a set of points**

house with 2000 square feet is 30 000 + 70(2000) = \$170 000. The intercept is the predicted value of the dependent variable when the independent variable is set equal to zero. In this example, a house with 0 square feet would sell at a predicted price of \$30 000! This intercept of \$30 000 could be interpreted as the value of the land on which the house is built. More generally, the intercept does not always have a realistic interpretation, since a zero value for the independent variable may lie well outside the range of observed values.

In studying the linear relationship between variables, each observation of the dependent variable, y, may be expressed as the sum of a predicted value and a residual term:

$$y = a + bx + e = \hat{y} + e \tag{6.3}$$

where $\hat{y} = a + bx$ is the predicted value, and e is termed the *residual*. The value $\hat{y}$ represents the value of the dependent variable predicted by the regression line. Note that the residual is equal to the difference between observed and predicted values:

$$e = y - \hat{y} \tag{6.4}$$

In keeping with the distinction between sample and population, note that a and b are estimates of some "true", unknown regression line. The slope and intercept of this true regression line could, in theory, be determined by taking a complete, 100% sample of the population. As usual, we use Greek letters to denote the population values of the parameters:

$$y = \alpha + \beta x + \varepsilon \tag{6.5}$$

where α and β are the intercept and slope of the true regression line, respectively. Each observation y may be viewed as the sum of a component that predicts the value of y on the basis of the value of x (using the true coefficients α and β) and some random error (ε). The error term reflects the fact that we do not expect the model to work "perfectly"; inevitably there will be other variables that also influence y, though we hope their influence is relatively minor. Observations on the dependent variable may be expressed as the sum of the predicted value and a "true" population error, ε, where

$$\varepsilon = y - \tilde{y} \tag{6.6}$$

is the difference between the observed value (y) and that predicted by the true regression line ($\tilde{y}$). The latter quantity is

$$\tilde{y} = \alpha + \beta x \tag{6.7}$$

In Figure 6.2 both the true regression line (Equation 6.7) and the best-fitting line based on the sample of points (Equation 6.1) are shown. It is important to

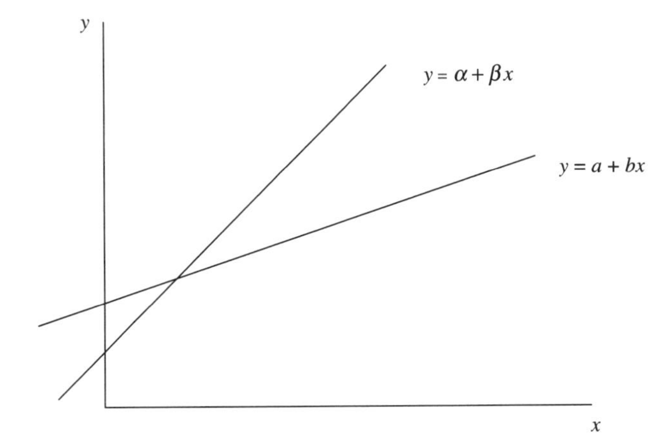

Figure 6.2 **True and sample regression lines**

keep in mind that, had a different sample been collected, the regression line based on the sample would be different but the true regression line would remain the same.

6.2 Fitting a Regression Line to a Set of Bivariate Data

Figure 6.3 shows a straight line fit through a set of points plotted in a two-dimensional, x–y space. In regression analysis, the objective is to find the slope and intercept of a best-fitting line that runs through the observed set of data points. But what is meant by best-fitting? There are certainly many ways to fit a line through a set of points. One way would be to fit the line so that the sum of the minimum distances of the observations to the line was a minimum. In this case, the distances are represented in Figure 6.3 geometrically as (dotted) lines that run from the observations to the regression line and which are perpendicular to the regression line.

In linear regression analysis, the sum of squared vertical distances from the observed points to the line (i.e., the solid lines in Figure 6.3) is minimized. The fact that vertical distances are used is consistent with the idea that the dependent variable, which is always portrayed on the vertical axis, is being predicted from the independent variable (portrayed on the horizontal axis). In fact, the vertical distance is identical to the value of the residual, which, as we have indicated, is the difference between the observed and predicted values of the dependent variable. Thus regression analysis minimizes the sum of squared residuals. The sum of *squared* residuals is used primarily for reasons of mathematical convenience – expressions for finding the values of a and b from the data are much easier to derive and express. Thus the objective is to find values of a and b that minimize the sum of squared residuals:

$$\min_{a,b} \sum_{i=1}^{n} (y_i - \hat{y})^2 = \min_{a,b} \sum_{i=1}^{n} (y_i - a - bx_i)^2 \tag{6.8}$$

Geometrically, this problem corresponds to finding the minimum of a three-dimensional parabolic cone, where a and b are coordinates in the two-dimensional plane and the sum of squared residuals is the vertical axis (see Figure 6.4). Viewing the figure, we can imagine trying different values of a and b – some will work relatively well in the sense that the sum of squared residuals will be quite small, and other combinations will be poor since the sum of squared residuals will be large. The values of a and b at the bottom of the parabolic cone may be determined using the data as follows (see inset for more detail):

$$\left.\begin{aligned} b &= \frac{\sum_{i=1}^{n}(x_i-\bar{x})(y_i-\bar{y})}{\sum_{i=1}^{n}(x_i-\bar{x})^2} \\ a &= \bar{y} - b\bar{x} \end{aligned}\right\} \tag{6.9}$$

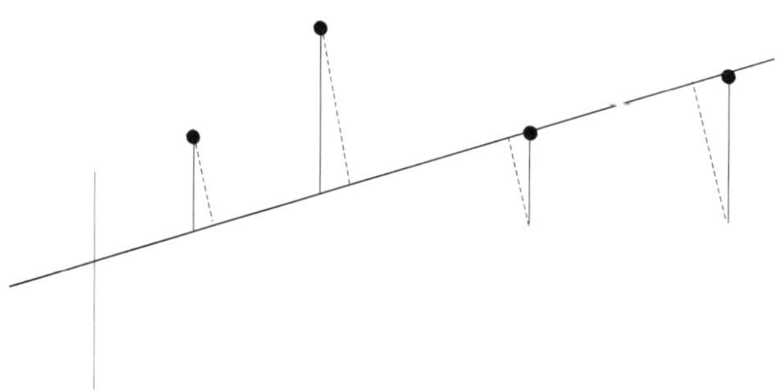

Figure 6.3 **Alternative measures of distance from points to regression line**

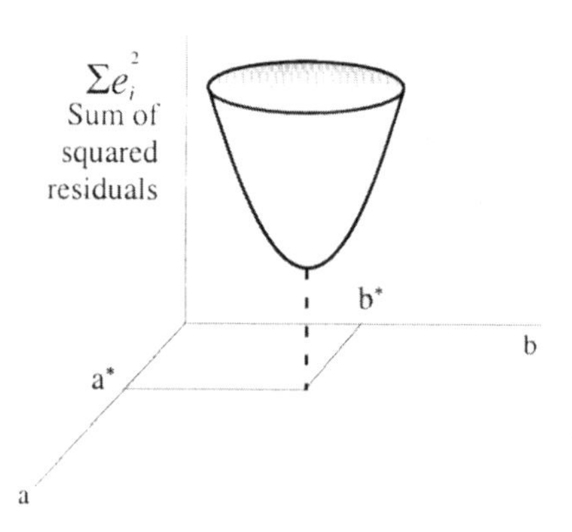

Figure 6.4 **Minimization of sum of squared residuals**

INSET: Finding the slope and intercept via the solution to a calculus problem
The solution of Equation 6.8 to find the values of the slope and intercept is the solution of a calculus problem. Solving the calculus problem amounts to finding the *a*, *b* combination that leads to the smallest sum of squared residuals at the bottom of the parabola. For those with a calculus background, we proceed by taking the derivatives of Equation 6.8 with respect to *a* and *b*, setting them each to zero, and solving for the two unknowns *a* and *b*. The result is Equation 6.9.

The reader should note that the numerator of the expression for the slope *b* is identical to that for the correlation coefficient *r*. In fact, for bivariate regression, the slope may be written in terms of the correlation coefficient:

$$b = r\frac{s_y}{s_x} \tag{6.10}$$

Once the values of *a* and *b* have been determined, plotting the line is straightforward. As indicated in the equation used above to determine *a*, the regression line goes through the mean $(\bar{x}, \bar{y})$. Another point on the line is $(0, a)$ (the intercept). After plotting these two points, the regression line may be drawn by connecting the two points together with a straight line.

6.3 Regression in Terms of Explained and Unexplained Sums of Squares

Another way to understand regression is to recognize that it provides a way to partition the variation in the observed values of a dependent (*y*) variable. In particular, the variability one observes in *y* may be decomposed into (a) a part that is explained by the regression line (via the assumption that *y* is linearly related to one or more independent variables) and (b) a part that remains unexplained.

More specifically, the variability in *y* may be measured by the sum of squared deviations of the *y* values from their mean. Some of this variability is explained by the regression line – the values of *y* vary partly because of the assumed relationship with the *x* variables. The partitioning of the total sum of squares into explained and unexplained components is analogous to the analysis of variance, where the sum of squares is divided into between- and within-column components. Here we have

$$\sum_{i=1}^{n} (y_i - \bar{y})^2 = \sum_{i=1}^{n} (\hat{y}_i - \bar{y})^2 + \sum_{i=1}^{n} (y_i - \hat{y})^2 \tag{6.11}$$

The left-hand side is the *total sum of squares*; the deviation between observed value and mean is represented geometrically by the distance *A* in Figure 6.5. In the figure, the points clearly vary about the horizontal line representing the mean value of *y*. Some of this variability may be attributed to the regression

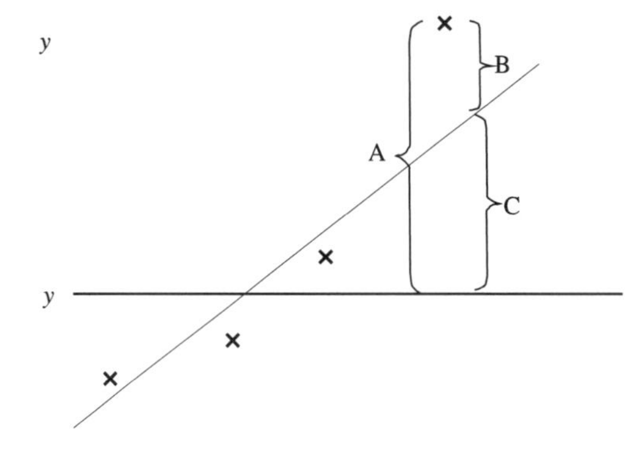

Figure 6.5 **Partitioning the variability in *y***

line; we expect points on the right-hand side of the diagram to be above the mean, and points on the left-side of the diagram to be below the mean.

The first term on the right-hand side is the *regression* (or *explained*) *sum of squares* – it is the sum of squared differences between predicted values and the mean value of y. These differences are the deviations of the regression line from the mean and constitute the "explained" portion of the variability in y. The distance between the predicted value and the mean is represented by C in Figure 6.5. Finally, the second term on the right-hand side is the *unexplained* or *residual sum of squares*. This is the sum of squared differences between the observed and predicted values. It is this quantity that is minimized when the coefficients are estimated. The distance between the observation and the regression line (i.e., the residual) is represented by B in the figure.

The proportion of the total variability in y explained by the regression is sometimes called the *coefficient of determination*, and it is equal to the square of the correlation coefficient:

$$r^2 = \frac{\sum_{i=1}^{n} (\hat{y}_i - \bar{y})^2}{\sum_{i=1}^{n} (y_i - \bar{y})^2} = 1 - \frac{\sum_{i=1}^{n} e_i^2}{(n-1)s_y^2} \tag{6.12}$$

where e is the value of the residual. Note that r^2 is equal to the regression sum of squares divided by the total sum of squares. It is also equal to one minus the ratio of the residual sum of squares to the total sum of squares.

The value of r^2 varies from 0 to 1; a value of zero would indicate that no variability has been explained, whereas a value of one would imply that all of the residuals are zero and the regression line fits perfectly through all of the observed points.

One way to determine whether the regression has been successful at explaining a significant portion of the variation in y is to perform an F-test, analogous to the F-test used in the analysis of variance. Specifically, for simple regression, the null hypothesis that $\rho^2 = 0$ is tested with the F-statistic

$$F = \frac{r^2(n-2)}{1-r^2} \tag{6.13}$$

which has an F-distribution with 1 and $n-2$ degrees of freedom when the null hypothesis is true. By looking back to the previous chapter on correlation, you will notice that this F-statistic is the square of the t-statistic used to test the hypothesis that the correlation coefficient ρ is equal to zero. The tests are identical in the sense that they will always yield identical conclusions and p-values.

The origins of this F-test lie in the partitioning of the sums of squares just described. We can create an analysis of variance table, for a hypothetical example with 12 observations, as follows:

	Sums of squares	df	Mean square	F
Regression (explained)	578	1	578	13.7
Residual (unexplained)	422	$n-2=10$	42.2	
Total	1000	$n-1=11$		

The total sum of squares has $n-1$ degrees of freedom associated with it. In simple regression, where there is one independent variable, there is always 1 degree of freedom associated with the regression sum of squares, leaving $n-2$ degrees of freedom associated with the residual sum of squares. The F-ratio is therefore equal to

$$F = \frac{\text{explained SS}}{(\text{residual SS})(n-2)} = \frac{578/1}{422/10} = 13.7 \tag{6.14}$$

Recalling the definition of r^2, this ANOVA-like expression for F can be seen to be equivalent to Equation 6.13, since

$$F = \frac{\text{explained SS}}{(\text{residual SS})/(n-2)} = \frac{r^2(n-2)}{1-r^2} \tag{6.15}$$

The value of r^2 may be thought of as the maximal correlation between a weighted combination of the independent variables and the dependent variable. (The weights happen to be the regression coefficients if the dependent and independent variables are put in their standardized, z-scores form). The calculated value of r^2 overestimates the true value, R^2. Note that if the number of observations is equal to the number of variables, r^2 will always equal one, even if the variables are unrelated (i.e., $R^2 = 0$). If $R^2 = 0$, then the expected value of r^2 is $(p-1)/(n-1)$, where p is the number of variables and n is the number of observations. For example, if $p = 11$ and $n = 21$, then the expected value of r^2 is $10/20 = 0.5$, even when the individual variables are not truly correlated! This serves to emphasize the importance of having a large number of observations relative to the number of variables. The *adjusted* r^2 represents a downward adjustment of r^2 that takes this difference between the sample and population values into account. We will see an example of this adjustment in Section 6.11.

6.4 Assumptions of Regression

The assumptions of regression analysis for simple regression are:

(1) The relationship between y and x is linear; that is, there is an equation $y = \alpha + \beta x + \varepsilon$ that constitutes the population model.
(2) The errors have mean zero and constant variance; that is, $E[\varepsilon] = 0$ and $V[\varepsilon] = \sigma^2$ The errors do not vary with x; that is, $V[\varepsilon | x] = \sigma_x^2 = \sigma^2$.
(3) The residuals are independent; the value of one error is not affected by the value of another error.
(4) For each value of x, the errors have a normal distribution about the regression line. This normal distribution is centered on the regression line. This assumption may be written $\varepsilon \sim N(0, \sigma^2)$.

Multiple regression, treated in the next chapter, adds another assumption – namely, that the x variables have no multicollinearity (that is, the independent variables are not significantly correlated with one another).

6.5 Standard Error of the Estimate

The standard error of the estimate is another expression for the standard deviation of the residuals; for the case of simple regression, it is estimated by

$$s_e = \sqrt{\sum_{i=1}^{n} \frac{(y_i - \hat{y}_i)^2}{n-2}} \tag{6.16}$$

6.6 Tests for Beta

We are often interested in testing the null hypothesis that the true value of the slope is equal to zero, i.e., $H_0 : \beta = 0$. We are, of course, usually interested in the prospect of rejecting this null hypothesis, thereby accumulating some evidence that the variable x is important in understanding y. This can be done via a t-test

$$t = \frac{b - \beta}{s_b} = \frac{b}{s_b} \tag{6.17}$$

where s_b is the standard deviation of the slope:

$$s_b = \sqrt{\frac{s_e^2}{(n-1)s_x^2}} \tag{6.18}$$

6.7 Confidence Intervals

A confidence interval for the regression line is found from

$$V[\hat{y}|x] = \frac{s_e^2}{n}\left(1 + \frac{(x - \bar{x})^2}{s_x^2}\right) \tag{6.19}$$

A confidence interval for individual predictions is found from

$$V[\hat{y}|x] = \frac{s_e^2}{n}\left(n + 1 + \frac{(x - \bar{x})^2}{s_x^2}\right) \tag{6.20}$$

6.8 Illustration: Income Levels and Consumer Expenditures

A supermarket is interested in how income levels (x) may affect the amount of money spent per week by its customers (y). The null hypothesis is that income levels do not affect the amount of money spent per week by customers, and the alternative hypothesis is that higher incomes are associated with greater spending. Table 6.1 depicts the data collected from ten survey respondents.

Table 6.1

Amount spent/week (y)	Income (×000) (x)
$120	65
$68	35
$35	30
$60	44
$100	80
$91	77
$44	32
$71	39
$89	44
$113	77

To fit the regression line, we can first compute the following quantities:

$$\left.\begin{array}{ll} \bar{x} = 52.3; & s_x = 20.20 \\ \bar{y} = 79.1; & s_y = 28.34 \\ \multicolumn{2}{l}{\sum_{i=1}^{n} (x_i - \bar{x})(y_i - \bar{y}) = 3672.1} \\ r = 0.835 & \end{array}\right\} \quad (6.21)$$

Income x is given in thousands of dollars, and weekly supermarket spending is given in dollars. From these, we may find the slope from either of the following:

$$\left.\begin{array}{l} b = r\dfrac{s_y}{s_x} = 0.835\dfrac{28.34}{20.2} = 1.171 \\ b = \dfrac{\sum_{i=1}^{n}(x_i - \bar{x})(y_i - \bar{y})}{\sum_{i=1}^{n}(x_i - \bar{x})^2} = \dfrac{4301.7}{20.2^2(9)} = 1.171 \end{array}\right\} \quad (6.22)$$

The intercept is

$$a = \bar{y} - b\bar{x} = 79.1 - 1.171(52.3) = 17.8 \quad (6.23)$$

Our regression line may be expressed as

$$\hat{y} = 17.8 + 1.171x \quad (6.24)$$

implying that every increase of \$1000 of income leads to an increase in \$1.17 spent at the supermarket each week.

For each observation, we can compute a predicted value and a residual, using the regression line, along with the values of x. For example, for the first observation, the predicted value of the dependent variable is

$$\hat{y} = 17.8 + (1.171)(65) = 93.9 \quad (6.25)$$

The residual for this observation is the observed value of y minus the predicted value:

$$e = 120 - 93.9 = 26.1 \quad (6.26)$$

Table 6.2 depicts the results, including residuals and predicted values for all observations. The sum of the residuals (subject to a bit of rounding error) is equal to zero – the amount by which the positive residuals lie above the regression line is equal to the amount by which the negative residuals lie below the regression line.

Table 6.2

Amount spent/week (y)	Income (×000) (x)	Predicted $\hat{y}$	Residual ε
$120	65	94.0	26.0
$68	35	58.8	9.2
$35	30	53.0	– 18.0
$60	44	69.4	– 9.4
$100	80	111.5	– 11.5
$91	77	108.0	– 17
$44	32	55.3	– 11.3
$71	39	63.5	7.5
$89	44	69.3	19.7
$113	77	108.0	5

The analysis of variance table associated with this regression is as follows:

	Sums of squares	df	Mean square	F
Regression (explained)	5039.2	1	5039.2	18.4
Residual (unexplained)	2189.7	$n-2=8$	273.7	
Total	7228.9	$n-1=9$		

This table can be constructed by recalling that, for bivariate regression, the number of degrees of freedom associated with the explained sum of squares is equal to one, and that associated with the total sum of squares is equal to $n-1$. Also recall that the mean square is simply the sum of squares divided by the number of degrees of freedom, and the F-statistic is the ratio of the sums of squares. The first column may be completed by recognizing that the total sum of squares is equal to the variance of y multiplied by $n-1$:

$$\text{Total sum of squares} = \sum_{i=1}^{n} (y_i - \bar{y})^2 = (n-1)s_y^2 \tag{6.27}$$

Since $r^2 = 0.835^2$ is the proportion of the total sum of squares explained by the regression, the regression sum of squares is equal to $7228.4(0.835^2) = 5039.8$. The residual sum of squares is simply the difference between the total and regression sums of squares.

The observed F-ratio of 18.4 in the table may be compared with the critical value of 5.32 found in the F-table, using $\alpha = 0.05$ and 1 and 8 degrees of freedom, respectively, for numerator and denominator. Since the observed value of F exceeds the critical value of 5.32, we reject the null hypothesis that the true correlation coefficient ρ^2 is equal to zero, and conclude that income explains a significant amount of the variability in supermarket spending.

We can also test the hypothesis that the true regression coefficient is equal to zero by using Equations 6.16 to 6.18. This first requires finding the variance of the residuals (s_e^2) and the standard error of the estimate (s_e):

$$s_e^2 = \frac{\sum_{i=1}^{n} e^2}{n-2} = \frac{2189.7}{8} = 273.7; \qquad s_e = \sqrt{273.7} = 16.54 \tag{6.28}$$

Next, the standard deviation of the estimate of the slope is given by

$$s_b = \sqrt{\frac{s_\varepsilon^2}{(n-1)s_x^2}} = \sqrt{\frac{273.7}{9(20.2^2)}} = \sqrt{0.0745} = 0.27 \tag{6.29}$$

The test of the null hypothesis $H_0 : \beta = 0$ is then carried out with the t-statistic

$$t = \frac{b - \beta}{s_b} = \frac{1.171}{0.27} = 4.34 \tag{6.30}$$

Since the observed value of t exceeds the critical value of 2.62 found from the table using a two-tailed test with $\alpha = 0.05$, $n - 1 = 9$ degrees of freedom, we reject the null hypothesis that the true regression coefficient is equal to zero. In bivariate regression, the t-test and the F-test will always give consistent results.

6.9 Illustration: State Aid to Secondary Schools

In New York State, aid is given to schools by the state government. The aid formula is based upon a number of factors, but one principle is that districts with residents that have relatively higher household incomes should receive relatively less in state aid.

A graph of state aid per pupil versus average household income for 27 public school districts in western New York reveals that there is in fact a downward trend in aid with increasing income (see Figure 6.6). Regression analysis of the 27 pairs of points confirms this:

$$\hat{y} = 6106.93 - 0.0818x \tag{6.31}$$

where $\hat{y}$ is the predicted amount of school aid per pupil, in dollars, and x is the average household income. The slope implies that every increase of \$1 in household income brings with it a decline in state aid of \$0.08 per pupil. Equivalently, every increase of \$1000 in household income leads to a decline of \$81.80 in state aid per pupil (although this estimate is also capturing the effects of other variables in the state aid formula that have been omitted here; see Sections 7.1.2 and 7.2). The value of r^2 is 0.257, and the slope is significantly negative (a t-test yields $t = -2.94$, which is more extreme than the critical value and implies a p-value of 0.007.

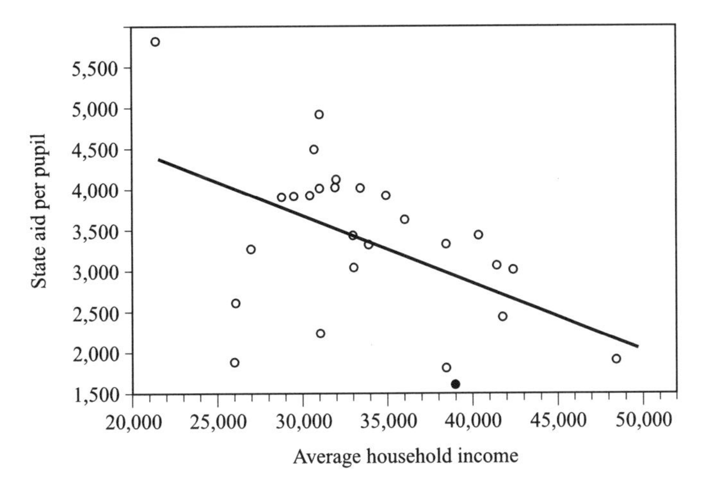

Figure 6.6 **New York State school aid per pupil vs average household income**

Of particular interest to me is the darkened circle, which represents the Amherst school district where my children attend school! Note that the state aid per pupil is significantly lower than the expected value, given the average household income in the district. A partial explanation of this has to do with the way in which the state collects information on income. All taxpayers must note their school district on their tax return, using a three-digit code taken from a list at the back of the tax form instruction booklet. The problem is that the Amherst school district lies in the town of Amherst, but so too do a number of other districts, including the Williamsville and Sweet Home districts. Residents of the Williamsville and Sweet Home districts dutifully go to the back of the instruction booklet, begin to scan the list, and quickly come across Amherst, since it is near the beginning of the alphabet. Since they live in the town of Amherst, they (incorrectly) copy the Amherst code onto their tax form. State officials then tally up the income in each district, and find (incorrectly) that Amherst residents make a lot of money and consequently should receive less in school aid! The average household income data used in Figure 6.6 represents the more accurate household income data taken from the US Census. Even though it may be out-of-date, since it was collected in 1990 and the analysis was done in 1997, the census data give a truer picture than do the tax return data of how deserving of aid each district is.

If the aid received by the Amherst school district from the state was in line with expectations, the black dot would be raised vertically from its present location to the regression line. This would represent an increase of \$1100 per pupil, which is almost a 70% increase over the present figure of about \$1600 per pupil. Since the district contains over 3000 students, this would represent an increase of over three million dollars. Unfortunately for the residents of the Amherst District, it is difficult to correct the imbalance, and this is true despite the potential of geographic information systems to attribute each resident's address to the correct local school district. There is a legal clause that limits the degree to which the problem can be corrected, since any change that gave more to the Amherst District would give less to surrounding districts. So, despite the fact that residents in the Williamsville and Sweet Home districts have less income attributed to their district than they should (and consequently have higher state aid), the imbalance has been only partially corrected.

One solution I suggested at a meeting of the Amherst School Board was that we change the name of our district! Since the problem is caused by the fact that the school district has the same name as the town, perhaps we could simply change the name to something else. This creative suggestion met with difficulty too, since it turns out that it is difficult, if not impossible, to change legally the school district's name. Another, less savory, approach would be to launch a campaign to encourage residents of the Amherst School district to put the school codes of Sweet Home or Williamsville on their tax forms. In any event, this example serves to illustrate how regression analysis can be used to both estimate the magnitude of effects of one variable on another (the term "effect" in the current example refers to the effect of income on state aid) and to interpret unusual observations. The example also illustrates that because of the quirky nature of data, potential pitfalls abound when one attempts to establish relationships between one variable (income) and another (state aid).

6.10 Linear versus Nonlinear Models

It should be understood that the word "linear" refers to the fact that in linear regression analysis the relationship is one that is *linear in the parameters*. The parameters are the intercept and slope coefficients, a and b. Thus linear regression could be used to study the effects of the square of some variable x on the value of y. Similarly, the equations

$$\left.\begin{aligned} y &= a + b\sqrt{x} \\ y &= a + b(\ln x)^2 \end{aligned}\right\} \tag{6.32}$$

may also be studied using the methods of linear regression, since the parameters a and b appear linearly (that is, they are raised to the power 1). One example of an equation that is *not* linear in its parameters is

$$y = a + b^2 x \tag{6.33}$$

since the parameter b is raised to the power 2.

In many instances, a nonlinear relationship may still be analyzed using linear regression, since the nonlinear curve may be transformed into a linear one. For example, a common finding in geographic research is that there is a distance–decay effect for many kinds of interaction. Furthermore, this effect is often well modeled with a negative exponential curve. Attendance at a local swimming pool, for instance, may appear to decline exponentially with the distance that individuals reside from the pool (see Figure 6.7). Negative exponential decay in this example could be modeled with an equation of the form

$$p_d = p_0 \mathrm{e}^{-bd} \tag{6.34}$$

where p_d is the pool attendance rate among residents residing a distance d from the pool and p_0 is the attendance rate among those living as close as possible to the pool. Furthermore, e is the constant 2.718..., and b is the rate at which attendance declines with distance (that is, it is a measure of the steepness of the exponential decline; higher values of b imply a greater effect of distance on attendance). In this example, we can transform the curvilinear relationship seen in Figure 6.2 into a linear one. One reason for wanting to do this is to be able to make use of the well-developed methods of linear regression analysis. The transformation is brought about by taking the logarithms of both sides of Equation 6.1:

$$\ln p_d = \ln p_0 - bd \tag{6.35}$$

This is the equation of a straight line; if the relation is linear, when $\ln p_d$ is plotted against distance d, the result will be a straight line with slope equal to b and intercept equal to $\ln p_0$ (see Figure 6.8).

Note also that models with negative exponential decline have to be initially written with a multiplicative error term, since that allows them to be linearized:

$$p_d = p_0 \mathrm{e}^{-bd} \varepsilon \Rightarrow \ln p_d = \ln p_0 - pd + \varepsilon \tag{6.36}$$

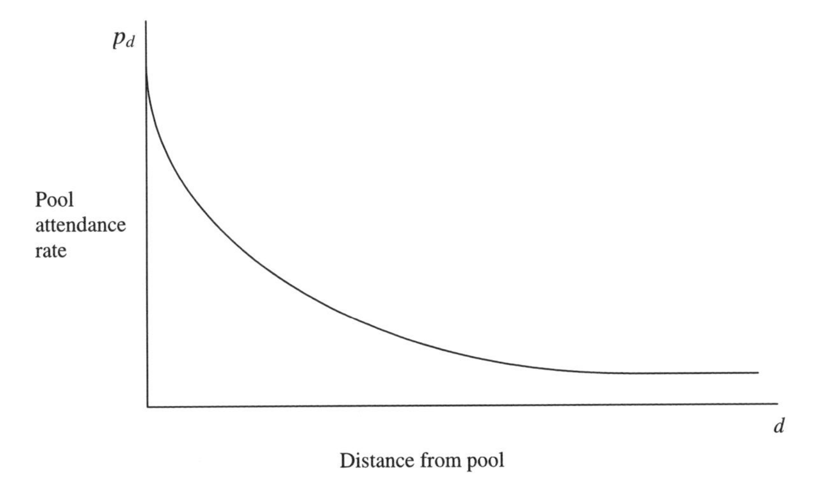

Figure 6.7 **Negative exponential decline in pool attendance rate with distance**

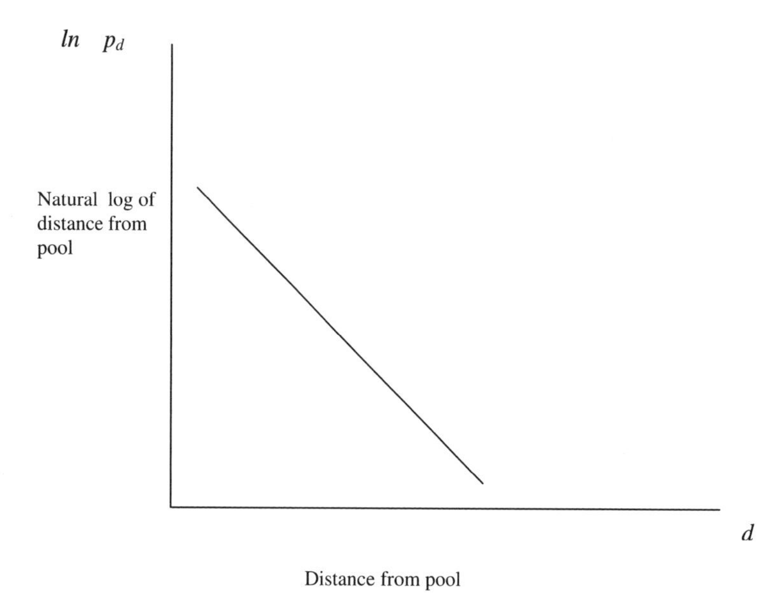

Figure 6.8 **Linear decline in log of pool attendance rate with distance**

If a model with negative exponential decline is written with an additive error term, as follows:

$$p_d = p_0 e^{-bd} + \varepsilon \quad (6.37)$$

it is said to be intrinsically nonlinear, since there is not a transformation that can convert it into the equation of a straight line.

In the next chapter, we will focus on multiple regression, where the regression model includes more than one explanatory variable. This leads to additional issues, and these are also discussed in Chapter 7.

6.11 Regression in *SPSS for Windows 9.0*

6.11.1 Data Input

Each observation is placed into a row of the data table. Each column of the data table corresponds to a variable. It is often convenient, but not necessary, to place the dependent variable in the first column.

6.11.2 Analysis

To carry out a regression analysis, first click on Analyze, then Regression, and then Linear; this will open a box. Within the box, select the dependent variable from the list of variables on the left, and use the arrow key to move it into the box titled "Dependent." Likewise, move the independent variables from the left to the box on the right labeled "Independent(s)." Clicking on "OK" will then carry out a regression analysis. This produces information on r, s_e, an ANOVA-type summary table, and information on the coefficients and their significance. Options for additional output are discussed below.

6.11.3 Options

Clicking on alternatives in the categories Statistics, Plots, and Save can produce additional output. It is common, for example, to want to save additional information. Under Save, one can click on boxes to save, among other items, predicted values, residuals, and confidence intervals associated with both the mean predicted value of y given x (i.e., the regression line; see Equation 6.19) and individual values of y (Equation 6.20). New columns containing the desired information are attached to the right-hand side of the data table.

Table 6.3 **Regression of amount spent per week vs income**

Variables Entered/Removed[b]

Model	Variables Entered	Variables Removed	Method
1	INCOME[a]	.	Enter

[a] All requested variables entered.
[b] Dependent Variable: AMTWEEK

Model Summary

Model	R	R Square	Adjusted R Square	Std. Error of the Estimate
1	.835[a]	.697	.659	16.5441

[a] Predictors: (Constant), INCOME

ANOVA[b]

Model		Sum of Squares	df	Mean Square	F	Sig.
1	Regression	5039.248	1	5039.248	18.411	.003[a]
	Residual	2189.652	8	273.706		
	Total	7228.900	9			

[a] Predictors: (Constant), INCOME
[b] Dependent Variable: AMTWEEK

Coefficients[a]

Model		Unstandardized Coefficients		Standardized Coefficients	t	Sig.
		B	Std. Error	Beta		
1	(Constant)	17.833	15.207		1.173	.275
	INCOME	1.171	.273	.835	4.291	.003

[a] Dependent Variable: AMTWEEK

6.11.4 Output

An example of output is shown in Table 6.3. This output corresponds to the results of the regression associated with the data in Table 6.1. Note that the value of r^2 is 0.697, and the adjusted r^2 value is 0.659.

Exercises

1. A regression of weekly shopping trip frequency on annual income (data entered in thousands of dollars) is performed on data collected from 24 respondents. The results are summarized below:

Intercept 0.46
Slope 0.19

	Sum of squares	df	Mean square	*F*
Regression				
Residual	1.7			
Total	2.3			

(a) Fill in the blanks in the ANOVA table.
(b) What is the predicted number of weekly shopping trips for someone making $50 000/ year?
(c) In words, what is the meaning of the coefficient 0.19?
(d) Is the regression coefficient significantly different from zero? How do you know?
(e) What is the value of the correlation coefficient?

2. Name four assumptions of simple linear regression.

3. The correlation coefficient and the slope are as follows:

$$r = \frac{\sum_{i=1}^{n}(x_i - \bar{x})(y_i - \bar{y})}{(n-1)s_x s_y}; \qquad b = \frac{\sum_{i=1}^{n}(x_i - \bar{x})(y_i - \bar{y})}{(n-1)s_x^2}$$

Find an equation for b in terms of r.

4. A regression of infant mortality rates (annual deaths per hundred births) on median annual household income (data entered in thousands of dollars) is performed on data collected from 34 counties. The results are summarized below:

Intercept 18.46
Slope −0.14

	Sum of squares	df	Mean square	*F*
Regression				
Residual	1.8			
Total	3.4			

(a) Fill in the blanks in the ANOVA table.
(b) What is the predicted infant mortality rate in a county where the median annual household income is $40 000?
(c) In words, what is the meaning of the coefficient -0.14? Do NOT simply say that this is the slope or the regression coefficient; indicate what it means and how it can be interpreted.
(d) What is the standard error of the residuals?
(e) Are predictions of mortality rates more accurate near the mean of the median incomes or away from the mean of median incomes?
(f) What is the value of the correlation coefficient?

5. A simple regression of Y vs X reveals, for $n=22$ observations, that $r^2=0.73$. The standard deviation of x is 2.3. The regression sum of squares is 1324. What is the value of the standard deviation of y? What is the value of the slope b?

6. In linear regression, which is wider, the confidence interval for a single predicted value of y, given x, or the confidence interval for the regression line? Give reasons for your answer.

7. Given a simple regression with slope $b=3$, $s_y=8$, and $s_x=2$, find the standard error of the estimate (i.e., the standard deviation of the residuals).

8. The following data are collected in an effort to determine whether snowfall is dependent upon elevation:

Snowfall (inches)	Elevation (feet)
36	400
78	800
11	200
45	675

Without the aid of a computer, show your work on problems (a) through (g).

(a) Find the regression coefficients (the intercept and the slope coefficient).
(b) Estimate the standard error of the residuals about the regression line.
(c) Test the hypothesis that the regression coefficient associated with the independent variable is equal to zero. Also place a 95% confidence interval on the regression coefficient.
(d) Find the value of R^2.
(e) Make a table of the observed values, predicted values, and residuals.
(f) Prepare an analysis of variance table portraying the regression results.
(g) Graph the data and the regression line.

7 More on Regression

LEARNING OBJECTIVES

- Regression with more than one independent explanatory variable
- Regression with categorical explanatory variables
- Regression with categorical dependent variables
- Interpreting multiple regression coefficients
- Choosing explanatory variables
- Consequences of poorly satisfied assumptions

7.1 Multiple Regression

It is most often the case that there is more than one variable that is thought to affect the dependent variable. For example, housing prices are affected by many characteristics of both the house and the neighborhood. The number of shopping trips generated by a residential neighborhood is affected by the income of its residents, the number of automobiles its residents own, accessibility to shopping alternatives, and so on.

With p independent explanatory variables, the regression equation is

$$\hat{y} = a + b_1x_1 + b_2x_2 + \cdots + b_px_p \tag{7.1}$$

where $\hat{y}$ is the predicted value of the dependent variable. With a given set of observations on the dependent (y) and independent (x) variables, the problem is to find the values of the parameters a and $b_1, b_2, \ldots, b_p$. The solution is found by minimizing the sum of the squared residuals:

$$\min_{\{a,b_1,\ldots,b_p\}} (y - a - b_1x_1 - \cdots - b_px_p)^2 \tag{7.2}$$

The problem and solution are identical in concept to that of bivariate regression discussed in the previous chapter, except that there are now more parameters to estimate and the geometric interpretation is carried out in a higher-dimensional space. If $p=2$, we wish to find a, b_1, and b_2 by fitting a plane through the set of points plotted in a three-dimensional space where the axes are represented by the y variable and the two x-variables (see Figure 7.1). The intercept a is the point of the plane on the y-axis when $x_1 = x_2 = 0$. The value of b_1 describes how much the value of y changes in the plane when x_1 increases by one unit along any line where x_2 is constant. Similarly, the value of

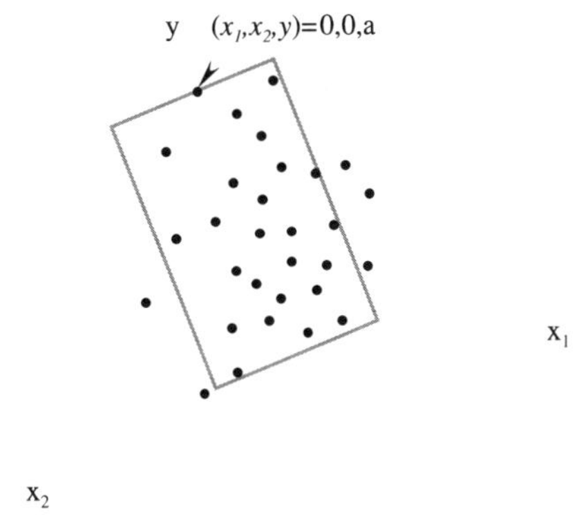

Figure 7.1 **Fitting a plane through a set of points in three dimensions**

b_2 describes the change in y when x_2 changes by one unit while x_1 is held constant. Although it is difficult (if not impossible!) to visualize, we wish to find the minimum of a four-dimensional parabolic cone, where the sum of squared residuals is represented along the vertical axis and the values of a, b_1, and b_2 are represented along the other dimensions.

More generally, for p independent variables, we fit a p-dimensional hyperplane through the set of points that are plotted in a $p+1$ dimensional space (one dimension for the y-variable and p additional dimensions, one for each of the independent variables). The coefficients a and $b_1, \ldots, b_p$ are found at the base of a $p+2$ dimensional parabola. Though it is of course not possible to actually picture this for high-dimensional spaces, this geometric description serves to reinforce what is actually being carried out in regression analysis.

Multiple regression carried out on spatial data raises special issues. One particularly difficult problem is that associated with the modifiable area unit problem, first discussed in Chapter 5. A regression of a dependent variable on a set of independent variables may yield substantially different conclusions when carried out on spatial units of differing sizes. Fotheringham and Wong (1991) note that with multiple and logistic regression (to be discussed in Section 7.6) the magnitude and significance of regression coefficients can be very sensitive to the size and configuration of areal units. If feasible, the sensitivity of one's results to changes in the size and/or shape of the spatial units should be explored.

7.1.1 Multicollinearity

In addition to the assumptions given for bivariate regression in the previous chapter, multiple regression analysis makes use of one additional assumption. In particular, it is assumed that there is no *multicollinearity* among the independent variables. This means that the correlation among the explanatory x-variables should not be high. In the extreme case where two variables are perfectly correlated, it is not possible to estimate the coefficients (and computer software will not provide results for this situation). In the more common case where multicollinearity is high, but not perfect, the estimates of the regression coefficients become very sensitive to individual observations; the addition or deletion of a few observations can change the coefficient estimates dramatically. Also, the variance of the coefficient estimates becomes inflated. Because the coefficients are more variable, it is not uncommon for insignificant independent, explanatory variables to appear significant.

7.1.2 Interpretation of Coefficients in Multiple Regression

Suppose that a regression of house prices in dollars (p) on lot size in square feet (x_1) and the number of bedrooms (x_2) results in the following equation:

$$p = 4000 + 20x_1 + 10\,000x_2 \qquad (7.3)$$

The coefficient on lot size means that every increase of one square foot adds an average of \$20 to the house price, *holding constant* the number of bedrooms. Similarly, the coefficient on the number of bedrooms implies that an added bedroom will increase the value of the house by an estimated \$10 000, for houses with identical lot sizes.

As with simple regression, the coefficients tell us the effect on the dependent variable of an increase of one unit in the independent variable. In addition, they control for the effects of other variables in the equation. That is, to understand the effect of a particular explanatory variable on the dependent variable, it is not sufficient to simply include it in the right-hand side of a regression equation. Since other variables may also affect the dependent variable, they also have to be included so that the separate effects of each contributing variable may be estimated. If all relevant variables are not included, this may lead to misspecification error.

7.2 Misspecification Error

Suppose that Equation 7.3 characterizes the "true" relationship between house prices, lot size, and the number of bedrooms. We will examine the effects of a misspecified regression equation that has an omitted variable by first making up some sample data from the equation

$$p = 4000 + 20x_1 + 10\,000x_2 + \varepsilon \qquad (7.4)$$

where ε is a normal random variable with mean 0 and variance equal to 3000^2 (this implies that one can, with 95% confidence, predict house prices within about two standard deviations, or \$6000). Table 7.1 displays the data associated with ten observations, where the data on lot size and the number

Table 7.1

House price	Lot size	Bedrooms
132 767	5000	3
134 689	5500	2
159 718	6000	4
164 937	6500	3
132 489	5200	2
125 766	5400	1
146 568	5700	3
168 932	6100	4
171 180	6300	4
187 921	6400	5

of bedrooms are simply hypothetical, and then Equation 7.4 is used to generate housing prices.

Now suppose that we incorrectly assume that house prices are a function of lot size only. A regression of house prices on lot size yields

$$\begin{array}{c} \hat{p} = -57\,809.7 + 36.2x_1 \\ (-1.7) \qquad (6.21) \end{array} \tag{7.5}$$

where the t-values associated with each coefficient are given below the equation in parentheses. We see that the coefficient of lot size is significant (since its t-value is greater than the one-sided critical value of $t=1.86$ with $n-2=8$ degrees of freedom) and is in the "right" direction (i.e., larger lot sizes lead to higher housing prices, as we would expect), but it is larger than the "true" value of 20. We have overestimated the effect of lot size by omitting the number of bedrooms, which also affects housing price.

Similarly, if we had incorrectly assumed that house prices were a function of the number of bedrooms only, we would find that the regression equation, based on the observed data, is

$$\begin{array}{c} \hat{p} = 103\,361 + 15\,580x_2 \\ (12.05) \qquad (6.1) \end{array} \tag{7.6}$$

The effect of number of bedrooms on housing price is significant and in the expected direction, but we have again overestimated somewhat the "true" effect of an added bedroom, which we know to be \$10 000.

Finally, if we use the data to estimate a regression equation with both independent variables, we find

$$\begin{array}{c} \hat{p} = -1993 + 21.6x_1 + 9333x_2 \\ (-0.13) \quad (7.12) \quad (7.24) \end{array} \tag{7.7}$$

Both variables have significant effects on housing prices, and, more importantly, we have estimated their effects on housing prices quite accurately, since the coefficient of 21.6 is close to its true value of 20 and the coefficient of 9333 is close to its true value of 10 000. The intercept is not too close to its true value of 4000, but we note from Table 7.2 that all of the true coefficients are within two standard deviations of their estimated values, and that all true values of the coefficients lie within their estimated confidence intervals.

7.3 Dummy Variables

It is sometimes necessary to include explanatory independent variables that are categorical. For instance, income is often not reported exactly but is rather classified into a category. Locations may be classified into, for example, central city, suburb, or rural categories.

To handle independent variables that have, say, k categories in regression analysis, we create $k - 1$ variables. One category is arbitrarily omitted; often the first category (e.g., lowest income) or last category (e.g., highest income) is omitted. Each of these new variables is a binary, 0–1 *dummy* variable. An observation is assigned a value of zero on one of these variables if it is *not* in the category, and is assigned a value of one if it *is* in the category.

Consider the example in Table 7.3, where individuals either report or are assigned one of three locations – central city, suburb, or rural. We will arbitrarily choose the rural region as the omitted category. We define $k - 1 = 2$ categories, the first associated with the central city and the second associated with the suburb.

The first two individuals live in the city – they are each assigned $x_1 = 1$ since they live in the city, and $x_2 = 0$ since they do not live in the suburb. Individuals 3 and 5 live in the suburb, so they are assigned $x_1 = 0$ since they do not live in the city, and $x_2 = 1$ since they live in the suburb. Note that individual 4 lives in the rural region, and is assigned a value of 0 on both x_1 and x_2, since the person does not live in either the city or the suburb.

Table 7.2

	Coefficient	Standard deviation	Confidence interval
Intercept	– 1993	14 881	(– 37 181, 33 195)
X_1	21.6	2.98	(14.6, 28.7)
X_2	9333	1310	(6234, 12 432)

Table 7.3

Individual	Location	x_1	x_2
1	Central city	1	0
2	Central city	1	0
3	Suburb	0	1
4	Rural	0	0
5	Suburb	0	1

The reason that a category is always omitted when dummy variables are employed has to do with multicollinearity. If all categories were included, there would be perfect multicollinearity, and this violates an assumption of multiple regression. Perfect multicollinearity would occur if we defined k dummy variables, since the sum of the k columns would always equal one. In our example above, we have included only two of the three categories; there is no reason to include a separate column for the third category, since it simply supplies us with redundant information (for example, we know that if individual 4 does not live in the central city or the suburb, he or she must live in the rural area).

Dummy variables are coded as 0 or 1 only, and not, for example, 1 or 2. The 0/1 coding is a result of the fact that the dummy variable is a nominal, categorical variable, and 0/1 coding corresponds to absence/presence.

Once dummy variables are defined, regression analysis proceeds in the usual way. Suppose that individuals one through five are observed to make 3, 4, 7, 1, and 5 weekly shopping trips, respectively. The resulting, best-fitting regression equation is found from a computer program to be

$$y = 1 + 2.5x_1 + 5x_2 \tag{7.8}$$

Table 7.4 displays the observed and predicted values.

The regression coefficients may be interpreted as follows. Being located in the rural region implies that x_1 and x_2 are zero, and so the predicted value of y is simply the intercept. Thus the intercept in dummy variable regression is the predicted value of the dependent variable for the omitted category. Being located in the central city is "worth" an extra 2.5 shopping trips, relative to the omitted category. Therefore, we predict that someone located in the central city will shop an average of 3.5 times per week. Being in the suburb is "worth" an extra 5 shopping trips per week, again, relative to the omitted (rural) category. Therefore, individuals residing in the suburb are predicted to shop $1+5=6$ times per week.

You may be wondering at this point what would have happened if we had omitted a category other than the rural region. Suppose the central city had been chosen as the omitted category. Then our data would look like those in Table 7.5. Here we define $x_1 = 1$ if the individual lives in the suburb, and $x_2 = 1$ if the individual lives in the rural region. Using this data in a multiple regression analysis yields

$$\hat{y} = 3.5 + 2.5x_1 - 2.5x_2 \tag{7.9}$$

Table 7.4 **Weekly shopping trip frequency**

Individual	Observed	Predicted
1	3	3.5
2	4	3.5
3	7	6
4	1	1
5	6	6

Table 7.5

Individual	Location	x_1	x_2	Weekly shopping trips
1	Central city	0	0	3
2	Central city	0	0	4
3	Suburb	1	0	7
4	Rural	0	1	1
5	Suburb	1	0	5

The coefficients are different, but when we interpret them in light of the new variable definitions, we come to the same conclusions as before (as we should!). For instance, the intercept of 3.5 is the predicted number of weekly shopping trips made by those in the central city (the omitted category). The coefficient of +2.5 is the extra number of shopping trips made by suburban residents relative to the omitted category. Thus suburban residents are predicted here to make $3.5+2.5=6$ shopping trips per week, the same as in the previous example. Similarly, the coefficient of -2.5 means that rural residents make 2.5 fewer shopping trips than central city residents do each week $(3.5-2.5=1.0)$.

7.3.1 Dummy Variable Regression in a Recreation Planning Example

Part of the statewide recreation planning process is to generate estimates and forecasts of recreation activity. Annual participation in a specific recreation activity is taken to be a function of variables such as age, income, and population density. In New York State, a survey of approximately 7500 people was undertaken; individuals were asked about their participation frequencies in various activities, and their age, income, and location were recorded. The independent explanatory variables were recorded as dummy variables. The dependent variable is the number of times the individual participated in the activity at organized public or private facilities, over the course of a year. A multiple regression analysis was run for each recreation activity. Table 7.6 displays the results.

For each independent variable, the highest category is omitted (i.e., high income, elderly, and urban locations). Recall that the coefficients are to be interpreted relative to these omitted categories. Thus a person in the low income category swims, on average, 5.55 times less per year than a person in the high income category. By using these coefficients, it is easy to estimate the participation frequency for any activity and any set of categories. For example, how often does a middle income, young adult, living in the rural regions of the state, participate in court games at organized facilities each year? The answer is 1.34 (the intercept, called the "base participation rate" in the table) plus 0.43 (the coefficient associated with those in the middle income category) plus 5.76 (for being a young adult) minus 3.65 (for those living in the rural area).

Table 7.6 **Recreation participation coefficients**

Activity	Base participation rate[a]	Income				Age				Population density		
		Low	Low–middle	Middle	High–middle	Youth	Young adult	Adult	Middle aged	Rural	Exurban	Suburban
Swimming	2.45	−5.55	−4.37	−0.73	3.22	22.83	9.83	8.94	2.91	8.51	5.98	4.78
Biking	−0.06	−0.94	−0.11	−0.92	0.07	21.63	6.21	3.82	1.17	1.64	2.95	1.31
Court games	1.34	−0.55	0.63	0.43	1.31	16.41	5.76	1.65	0.12	−3.65	−2.11	−1.54
Camping	0.27	−0.13	−0.01	0.34	0.39	1.93	0.44	−0.01	−0.19	1.25	0.41	0.80
Tennis	0.74	−2.30	−2.42	−2.45	−0.46	7.76	4.28	3.31	0.61	1.48	1.78	2.14
Picnicking	0.66	−0.15	0.31	0.48	0.67	1.87	2.23	1.84	0.67	2.30	0.97	1.10
Golf	0.93	−1.44	−1.38	−0.71	−0.58	−0.30	−0.30	0.28	0.70	1.02	1.34	0.42
Fishing	−0.21	1.35	0.17	0.57	0.83	3.36	1.26	2.01	0.72	1.77	1.41	1.17
Hiking	1.31	0.22	0.20	−0.04	0.77	2.49	0.65	0.47	0.18	0.53	0.54	0.30
Boating	0.24	−1.06	−0.77	−0.09	0.45	3.58	1.42	1.28	0.50	2.36	1.68	1.86
Field games	−0.17	−0.39	−0.19	1.53	0.86	8.22	2.73	0.94	−0.13	1.48	0.83	0.56
Skiing	0.70	−0.98	−1.10	−0.86	−0.36	0.85	0.37	0.63	0.03	0.19	0.78	0.31
Snowmobiling	−0.07	−0.50	−0.48	−0.08	−0.37	0.33	0.85	0.20	0.14	1.52	0.97	0.44
Local winter	0.66	−1.49	−1.18	−0.80	0.15	5.13	1.34	0.82	0.06	1.84	−0.32	1.05

[a] The control group selected was the highest income, age, and density group. This column gives the estimated base annual participation rate for this group. The other columns give the amounts to be added to this amount to obtain participation rates for any other income, age, and density group. Negative participation rates are possible and should be interpreted as zero.

Source: New York State Office of Parks and Recreation (1978).

Our estimate is therefore equal to 3.88 times per year. It should of course be kept in mind that this is an *average* participation rate across all individuals in that category.

If an individual happens to be in an omitted category, the implied coefficient is equal to zero. Thus high–middle income elderly residents of urban areas swim on average 2.45 (base participation rate, or intercept) plus 3.22 (income coefficient) plus 0 (since they are in the omitted, elderly category) plus 0 (since they are also in the omitted, urban category), which is equal to 5.67 times per year.

Recreation planners use these coefficients to plan for future use. Demographic projections provide forecasts of the number of people in each age/income/population density category. If we can rely on the coefficients in the table as estimates of individual recreation participation, we can use the coefficients together with the demographic forecasts to project how much demand there will be for each recreation activity. The state can then prioritize recreation projects in a way that will best meet the anticipated demand.

7.4 Multiple Regression Illustration: Species in the Galapagos Islands

The data in Table 7.7 contain two possible dependent variables related to the number of species found on 30 islands (total species, and number of native species), as well as five potential independent variables that may help to understand why different islands have different numbers of species.

In our example, we will use the total number of species as the dependent variable. We will now explore some of the choices and questions that one is faced with in arriving at a suitable regression model. All of the output shown in the tables is from *SPSS for Windows 9.0*.

7.4.1 Model 1: The Kitchen-Sink Approach

One idea would be to simply put all five independent variables on the right-hand side and see what happens – i.e., everything is put into the equation except the kitchen sink! This approach is *not* recommended, and is shown here for illustrative purposes only. The output in Table 7.8 shows the value of r^2 to be 0.877. There are two significant variables – elevation has a positive effect on species number, and the area of the adjacent island has a negative effect. Note that the sign of the area variable is negative, which is counter to one's intuition that more species would be found on larger islands. The standard error of the estimate is 66, which is approximately equivalent to the average absolute value of a residual (in this case the actual average absolute value of a residual is 44). This is pretty high, since half of the islands have fewer than 44 species! Finally, note that the ANOVA table is similar to that in the univariate case, with the regression degrees of freedom equal to the number of independent variables.

It is often tempting to use the kitchen-sink approach because when an independent variable is added to the regression equation, the R^2 value *always* increases. It is important to realize that a high R^2 value is *not* the primary goal of regression analysis; if it were, we could simply keep adding explanatory variables until we achieved our desired value of R^2! A more reasonable strategy often involves either (a) deleting variables that do not reduce the value of R^2 very much, and/or (b) adding variables only when they increase R^2 appreciably. Variable selection is discussed in more detail in Section 7.5.

Table 7.7 **Galapagos Islands: species and geography**

Island	Observed species		Area	Elevation	Distance (km)		Area of adjacent island
	Number	Native	km^2	(m)	From nearest island	From Santa Cruz	(km^2)
1 Baltra	58	23	25.09	–	0.6	0.6	1.84
2 Bartolomé	31	21	1.24	109	0.6	26.3	572.33
3 Caldwell	3	3	0.21	114	2.8	58.7	0.78
4 Champion	25	9	0.10	46	1.9	47.4	0.18
5 Coamaño	2	1	0.05	–	1.9	1.9	903.82
6 Daphne Major	18	11	0.34	119	8.0	8.0	1.84
7 Daphne Minor	24	–	0.08	93	6.0	12.0	0.34
8 Darwin	10	7	2.33	168	34.1	290.2	2.85
9 Eden	8	4	0.03	–	0.4	0.4	17.95
10 Enderby	2	2	0.18	112	2.6	50.2	0.10
11 Espanola	97	26	58.27	198	1.1	88.3	0.57
12 Fernandina	93	35	634.49	1494	4.3	95.3	4669.32
13 Gardner*	58	17	0.57	49	1.1	93.1	58.27
14 Gardner†	5	4	0.78	227	4.6	62.2	0.21
15 Genovesa	40	19	17.35	76	47.4	92.2	129.49
16 Isabela	347	89	4669.32	1707	0.7	28.1	634.49
17 Marchena	51	23	129.49	343	29.1	85.9	59.56
18 Onslow	2	2	0.01	25	3.3	45.9	0.10
19 Pinta	104	37	59.56	777	29.1	119.6	129.49
20 Pinzón	108	33	17.95	458	10.7	10.7	0.03
21 Las Plazas	12	9	0.23	–	0.5	0.6	25.09
22 Rabida	70	30	4.89	367	4.4	24.4	572.33
23 San Cristóbal	280	65	551.62	716	45.2	66.6	0.57
24 San Salvador	237	81	572.33	906	0.2	19.8	4.89
25 Santa Cruz	444	95	903.82	864	0.6	0.0	0.52
26 Santa Fé	62	28	24.08	259	16.5	16.5	0.52
27 Santa Maria	285	73	170.92	640	2.6	49.2	0.10
28 Seymour	44	16	1.84	–	0.6	9.6	25.09
29 Tortuga	16	8	1.24	186	6.8	50.9	17.95
30 Wolf	21	12	2.85	253	34.1	254.7	2.33

*Near Espanola. † Near Santa Maria.
The values marked – are not known.

Source: Andrews and Herzberg (1985).

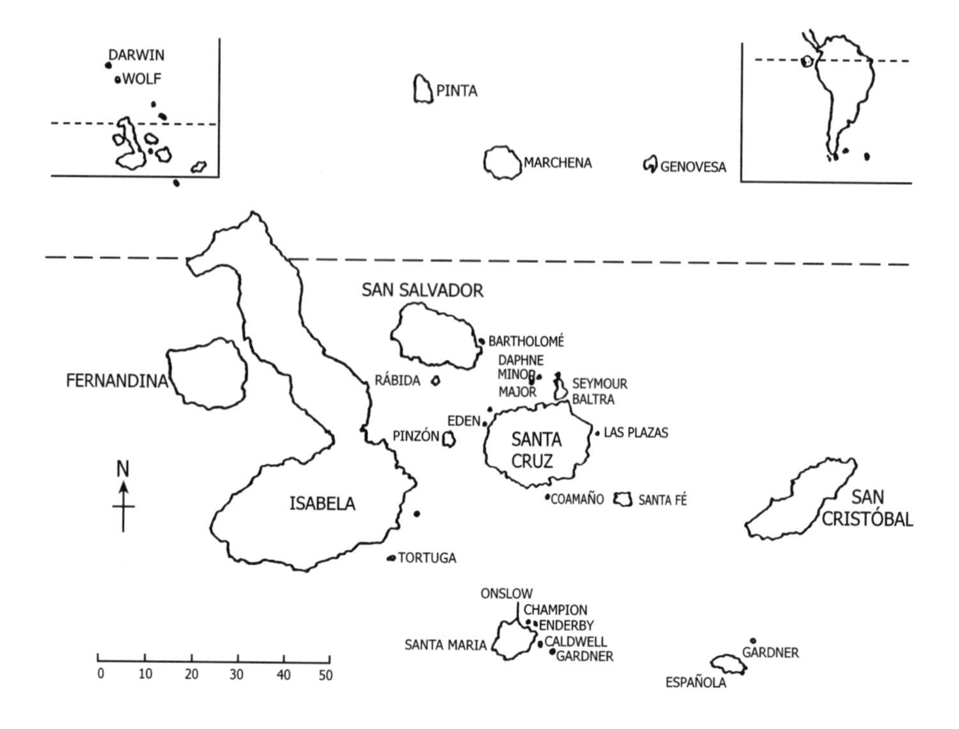

7.4.2 Missing values

Delving right into the analysis without a consideration of certain questions can lead to misinterpretation. Before we start, we should decide how we are going to treat missing values. In the table there are five missing values of elevation. All of the missing values, with the exception of the first observation, are for islands that have extremely small areas (and, with the exception of Seymour, small numbers of species). There are various ways we can proceed, including:

(1) Delete all observations with missing values. This is the default option used by many software packages.
(2) Replace missing values with the mean. Most statistical software packages have an option that allows missing values to be replaced with the mean of the remaining values.
(3) Use the other independent variables, or some subset of them, to predict the missing value. We could perform an initial regression of elevation on the other four independent variables for the nonmissing cases. Then we can use the results to predict the value of elevation, based upon the values of the other independent variables, for the missing observations.

Which option, or combination of options, should we choose? Option two is not a reasonable one here. The mean elevation is 412 m, and it would be

Table 7.8 **The kitchen-sink model**

Variables Entered/Removed[b]

Model	Variables Entered	Variables Removed	Method
1	AREAADJ, DISSC, AREA, DISNISL, ELEV[a]		Enter

[a] All requested variables entered.
[b] Dependent Variable: SPECIES

Model Summary

Model	R	R Square	Adjusted R Square	Std. Error of the Estimate
1	.877[a]	.768	.707	65.9482

[a] Predictors: (Constant), AREAADJ, DISSC, AREA, DISNISL, ELEV

ANOVA[b]

Model		Sum of Squares	Df	Mean Square	F	Sig.
1	Regression	274097.4	5	54819.479	12.605	.000[a]
	Residual	82634.046	19	4349.160		
	Total	356.731.4	24			

[a] Predictors: (Constant), AREAADJ, DISSC, AREA, DISNISL, ELEV
[b] Dependent Variable: SPECIES

Coefficients[a]

Model		Unstandardized Coefficients		Standardized Coefficients	t	Sig.
		B	Std. Error	Beta		
1	(Constant)	11.485	25.626		.448	.659
	AREA	-2.85E-02	.025	-.220	-1.141	.268
	ELEV	.330	.063	1.212	5.270	.000
	DISNISL	-.155	1.149	-.019	-.134	.894
	DISSC	-.260	.243	-.149	-1.068	.299
	AREAADJ	-7.96E-02	.020	-.611	-3.903	.001

[a] Dependent Variable: SPECIES

foolish to suppose that the unknown elevations are this great on the very small islands for which we have no data.

For the very small islands (observations 5, 9, and 21), one can justifiably exclude them from the analysis. Although we might also exclude Baltra and Seymour, here we will estimate their elevations from a regression of elevation on area. The resulting regression equation is

$$\text{Elevation} = 300 + 0.358(\text{Area}) \qquad (7.10)$$

We estimate Baltra's elevation as $300 + 25.09(0.358) = 309$ m, and Seymour's as $300 + 1.84(0.358) = 301$ m.

7.4.3 Outliers and Multicollinearity

A cursory examination of the data reveals that there is a small number of large islands. This is not an unusual feature of many studies, and it is important that we know whether these observations are exerting a significant effect on the results. *Leverage values* are designed to indicate how influential particular observations are in regression analysis. If the leverage value exceeds $2p/n$, where p is the number of independent variables, the observation should be considered as an outlier.

We also want to make sure that multicollinearity is not exerting an undue influence on the results. An examination of the correlations among the independent variables will reveal those where the high correlations exist. The *tolerance* is equal to the amount of variance in an independent variable that is not explained by the other independent variables. It is equal to $1 - r^2$, where the r^2 is associated with the regression of the independent variable on all other independent variables. A low tolerance indicates problems with multicollinearity, since the variable in question has a high correlation with the other independent variables. The reciprocal of the tolerance is the *variance inflation factor* (VIF); if it is greater than about 5, this indicates potential multicollinearity problems.

7.4.4 Model 2

In this second model, we have accounted for the missing elevation data and have output information on outliers and multicollinearity.

From the output (Table 7.9), we see that the inclusion of Baltra and Seymour has not changed the results very much. The r^2 value is 0.869, the standard error of the residuals is about 65, and elevation and the area of the adjacent island are still the only significant independent variables. We learn, however, that the variance inflation factor is slightly high (though not greater than the rule-of-thumb value of 5) for elevation and area. This is not surprising when we also inspect the correlation matrix, which reveals a very high correlation between the two variables.

One option for treating multicollinearity is to exclude from the analysis one or more variables. Here area and elevation are correlated, and we might decide to drop one of the two since they are close to being redundant (and since the sign of one of them is not correct). Which one should we drop? The choice should come primarily from a consideration of the underlying process and, secondarily, from the magnitude of the variance inflation factors. Both area and elevation should affect species number, but we will choose to exclude area because (a) elevation is important in terms of species diversity, and (b) the VIF is slightly higher for area than it is for elevation. A note of caution is in order here. Dropping variables from the analysis should only occur after a well-reasoned consideration of the underlying process. It does little to advance

Table 7.9 **Regression estimation with outliers removed**

Variables Entered/Removed[b]

Model	Variables Entered	Variables Removed	Method
1	AREAADJ, DISSC, AREA, DISNISL, ELEV[a]		Enter

[a] All requested variables entered.
[b] Dependent Variable: SPECIES

Model Summary[b]

Model	R	R Square	Adjusted R Square	Std. Error of the Estimate
1	.869[a]	.755	.697	64.8830

[a] Predictors: (Constant), AREAADJ, DISSC, AREA, DISNISL, ELEV
[b] Dependent Variable: SPECIES

ANOVA[b]

Model		Sum of Squares	df	Mean Square	F	Sig.
1	Regression	272396.7	5	54479.338	12.941	.000[a]
	Residual	88405.975	21	4209.808		
	Total	360802.7	26			

[a] Predictors: (Constant), AREAADJ, DISSC, AREA, DISNISL, ELEV
[b] Dependent Variable: SPECIES

Coefficients[a]

Model		Unstandardized Coefficients		Standardized Coefficients	t	Sig.	Collinearity Statistics	
		B	Std.Error	Beta			Tolerance	VIF
1	(Constant)	3.501	24.263		.144	.887		
	AREA	−2.50E-02	.024	−.192	−1.023	.318	.331	3.026
	ELEV	.325	.061	1.191	5.291	.000	.230	4.339
	DISNISL	−7.93E-03	1.124	−.001	−.007	.994	.592	1.689
	DISSC	−222	.237	−.130	−.936	.360	.604	1.654
	AREAADJ	−7.76E-02	.020	−.594	−3.880	.001	.498	2.009

[a]. Dependent Variable: SPECIES

Collinearity Diagnostics[a]

Model	Dimension	Eigenvalue	Condition Index	Variance Proportions				
				(Constant)	AREA	ELEV	DISNISL	DISSC
1	1	3.170	1.000	.02	.01	.01	.02	.02
	2	1.416	1.496	.00	.06	.01	.06	.04
	3	.764	2.037	.00	.12	.00	.00	.00
	4	.347	3.021	.45	.09	.00	.23	.05
	5	.231	3.702	.00	.05	.02	.59	.77
	6	7.222E-02	6.625	.52	.67	.95	.10	.12

[a] Dependent Variable: SPECIES

understanding of the process when important variables are dropped from the regression equation solely because they don't perform well.

The leverage values also reveal that several outliers have an important impact upon the results. Leverage values are over the rule-of-thumb value of $2p/n = 10/27 = 0.37$ for observations 8, 12, 15, and 16. Fernandina (observation 12) and Isabela (observation 16) have by far the two highest elevations among the thirty islands. There seems to be less justification for deleting the other two observations. Darwin (observation 8) and Genovesa (15) are geographic outliers, but there are also other geographic outliers with low leverage values.

7.4.5 Model 3

In this third model, we have deleted area as an independent variable and have deleted the observations for Fernandina and Isabela. The output (Table 7.10)

Table 7.10 **Regression with missing data removed or estimated, and outliers and area variable removed**

Variables Entered/Removed[b]

Model	Variables Entered	Variables Removed	Method
1	AREAADJ, ELEV, DISNISL, DISSC[a]		Enter

[a] All requested variables entered.
[b] Dependent Variable: SPECIES

Model Summary[b]

Model	R	R Square	Adjusted R Square	Std. Error of the Estimate
1	.858[a]	.736	.684	62.2581

[a] Predictors: (Constant), AREAADJ, ELEV, DISNISL, DISSC
[b] Dependent Variable: SPECIES

ANOVA[b]

Model		Sum of Squares	df	Mean Square	F	Sig.
1	Regression	216670.7	4	54167.666	13.975	.000[a]
	Residual	77521.336	20	3876.067		
	Total	294192.0	24			

[a] Predictors: (Constant), AREAADJ, ELEV, DISNISL, DISSC
[b] Dependent Variable: SPECIES

Coefficients[a]

Model		Unstandardized Coefficients		Standardized Coefficients	t	Sig.	Collinearity Statistics	
		B	Std.Error	Beta			Tolerance	VIF
1	(Constant)	-5.179	24.690		-.210	.836		
	ELEV	.344	.050	.831	6.936	.000	.919	1.088
	DISNISL	-.267	1.086	-.036	-.246	.808	.602	1.662
	DISSC	-.170	.232	-.109	-.730	.474	.591	1.691
	AREAADJ	-5.15E-02	.081	-.073	-.632	.534	.981	1.020

[a] Dependent Variable: SPECIES

Collinearity Diagnostics[a]

Model	Dimension	Eigenvalue	Condition Index	Variance Proportions				
				(Constant)	ELEV	DISNISL	DISSC	AREAAD
1	1	3.030	1.000	.02	.03	.03	.02	.02
	2	.932	1.804	.00	.00	.03	.03	.72
	3	.618	2.214	.03	.32	.07	.10	.16
	4	.274	3.325	.22	.09	.64	.25	.03
	5	.146	4.556	.73	.56	.23	.59	.07

[a] Dependent Variable: SPECIES

shows that only elevation remains significant. The value of r^2 remains high at 0.858, and the standard error of the estimate is slightly lower, at about 62. Multicollinearity is not an issue, since all of the VIFs are less than 5. Three observations (Bartolome, Darwin, and Rabida) are still outliers. This is likely due to their extreme values on some of the independent variables. Bartolomé and Rabida are both adjacent to large islands, and Darwin is a long way from Santa Cruz.

7.4.6 Model 4

To end up with a parsimonious model, we can remove the variables that are not significant. Thus we regress species number on elevation. The result is the equation

$$\text{Species} = -24.27 + 0.35(\text{Elevation}) \qquad (7.11)$$

From Table 7.11, the value of r^2 is still high at 0.846 (it is necessarily lower than before since we have removed variables, but it has not declined very much). The standard error of the estimate is about 60. Furthermore, a check of the leverage values reveals no outliers.

Table 7.11 **Final regression equation for species data**

Variables Entered/Removed[b]

Model	Variables Entered	Variables Removed	Method
1	ELEV[a]		Enter

[a] All requested variables entered.
[b] Dependent Variable: SPECIES

Model Summary[b]

Model	R	R Square	Adjusted R Square	Std. Error of the Estimate
1	.846[a]	.716	.703	60.3225

ANOVA[b]

Model		Sum of Squares	df	Mean Square	F	Sig
1	Regression	210499.4	1	210499.4	57.848	.000[a]
	Residual	83692.624	23	3638.810		
	Total	294192.0	24			

Coefficients[a]

Model		Unstandardized Coefficients		Standardized Coefficients		
		B	Std.Error	Beta	t	Sig
1	(Constant)	−24.270	18.640		−1.302	.206
	ELEV	.350	.046	.846	7.606	.000

[a] Dependent Variable: SPECIES

7.5 Variable Selection

As seen in the example, a common issue in regression analysis is the selection of variables that will appear as explanatory variables on the right-hand side of the regression equation. A brute-force approach to this question would be to try all possible combinations. With p potential independent variables this would mean that we would try p separate regressions that have just one independent variable, all $\binom{p}{2} = p(p-1)/2$ equations that have two variables, all $\binom{p}{3} = p(p-1)(p-2)/6$ equations that have three variables, and so on. If p is large, this is a lot of equations. Even with $p=5$, one would have to try 31 regression equations. But perhaps the real drawback is that this is even more of a "kitchen-sink" approach than to include all of the variables on the right-hand side. It amounts to an admission that we don't know what we are doing, and that our strategy is just to go with what looks best. It is always desirable to start from hypotheses and underlying processes first, in keeping with the principles of the scientific method described in the first chapter. In a more exploratory spirit, however, there may be some cases where we really have little in the way of *a priori* hypotheses – in this case, the all-possible regressions approach might be viewed as a potential way to generate new hypotheses.

An alternative way to select variables for inclusion in a regression equation is the *forward selection* approach. The variable that is most highly correlated with the dependent variable is entered first. Then, given that that variable is already in the equation, a search is made to see whether there are other variables that would be significant if added. If so, the one with the greatest significance is added. In this way, a regression equation is built up. The procedure terminates when there are no variables in the set of potential variables that would be significant if entered into the equation.

Backward selection starts with the kitchen-sink equation, where all of the possible independent variables are in the equation. Then the one that contributes least to the r^2 value is removed if the reduction in r^2 is not significant. The process of removing variables continues until the removal of any variable in the equation would constitute a significant reduction in r^2.

Stepwise regression is a combination of the forward and backward procedures. Variables are added in the manner of forward selection. However, as each variable is added, variables entered on earlier steps are re-checked to see if they are still significant. If they are not still significant, they are removed.

7.6 Categorical Dependent Variable

There are many situations where the dependent variable will be a categorical variable. For example, we may wish to model whether individuals patronize a park as a function of the distance to the park, or whether individuals commute by train as a function of automobile travel time. We may want to estimate the probability that a customer patronizes any of, say, four supermarkets as a function of characteristics of the stores and characteristics of the customers.

In each of these examples, the dependent variable is categorical in the sense that possible outcomes may be placed into categories. An individual either goes to the park or does not go. An individual either commutes by train or does not. If there are only four choices of supermarkets in an area, the consumers may be classified according to which one they patronize.

When the dependent variable is categorical, special consideration must be given to how regression analyses are carried out. In this section, we will examine why this is the case, and we will find out how *logistic regression* may be used in such situations.

7.6.1 Binary Response

In the simplest case, there are two possible responses. For example, we may assign the dependent variable a value of $y=1$ if the individual takes the train to work, and $y=0$ otherwise. Suppose, for instance, we had the data in Table 7.12 for $n=12$ respondents.

A cursory examination of the table reveals that there seems to be a tendency to take the train when the travel time by automobile would be high. Where auto travel time is not as high, there is more of a tendency for the y variable to be zero, indicating that the individual drives to work.

We could begin by running an ordinary least-squares analysis; we would find

$$\hat{y} = -0.396 + 0.0153x \tag{7.12}$$

The value of $\hat{y}$, which is a continuous variable, may be interpreted as the predicted probability of taking the train, given an automobile travel time equal to x.

There are several problems with this approach. One is that the assumption of homoscedasticity is not met; the estimated variance about the regression line is

Table 7.12

y	x: Auto travel time (min)
0	32
1	89
0	50
1	49
0	80
1	56
0	40
1	70
1	72
1	76
0	32
0	58

equal to $y(1-y)$ and therefore is not constant. Perhaps more troubling is that the predicted probabilities ($\hat{y}$) do not have to stay on the $(0, 1)$ interval; the reader might confirm that values of x less than around 25 will yield negative probabilities, and values of y greater than about 100 will yield probabilities greater than one! The problem is as shown in Figure 7.2.

How then should we proceed? One idea is to make the probabilities relate to x in a nonlinear way. The logistic equation in Figure 7.3 has the following equation:

$$\hat{y}_i = \frac{e^{\alpha+\beta x_i}}{1+e^{\alpha+\beta x_i}} \tag{7.13}$$

Note that when $\alpha + \beta x$ is a large negative number the predicted probability is near zero, whereas if $\alpha + \beta x$ is a large positive number the predicted

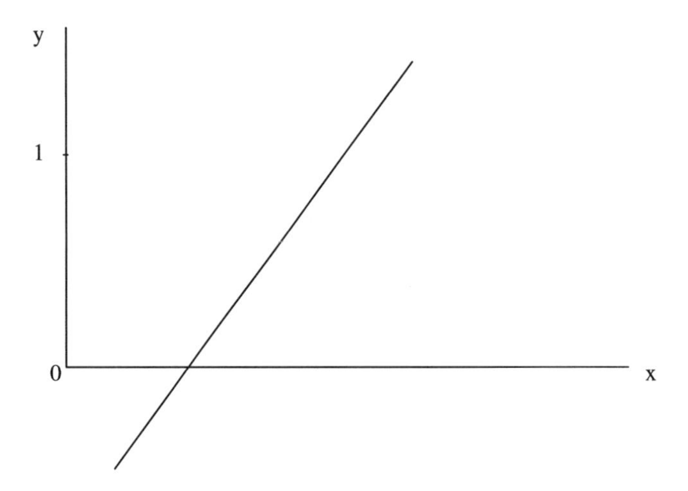

Figure 7.2 **Predicted probabilities outside the (0,1) interval**

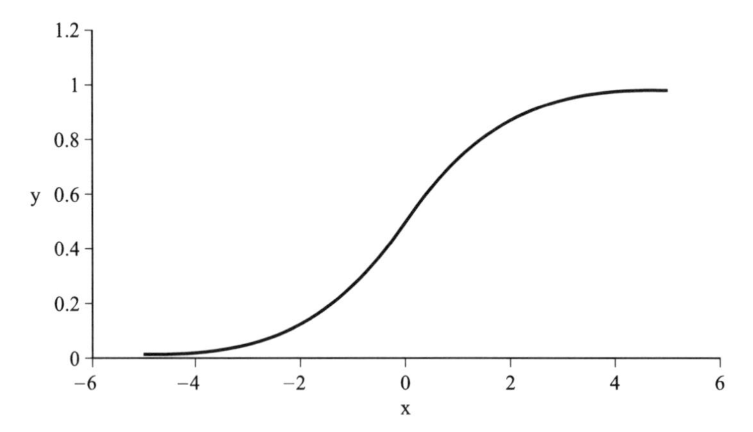

Figure 7.3 **The logistic curve**

probability is near one. In these cases, the predicted probabilities approach their asymptotes of 0 or 1 but never actually reach them. Thus it is no longer possible to predict probabilities that are either negative or greater than one.

While we've solved one problem by keeping the dependent variable on the (0, 1) interval, we've created another. How can we estimate the parameters? We can't use linear regression, since the equation is clearly not linear. One approach is to use nonlinear least squares. Specifically, we want to find α and β to minimize the sum of squared deviations between observed and predicted values

$$\min \sum_{i=1}^{n} (y_i - \hat{y}_i)^2 \tag{7.14}$$

where the predicted value, $\hat{y}_i$, is given by Equation 7.13 above. The answer, when minimizing the sum of squared residuals for the data in Table 7.12, is $\alpha = -4.501$ and $\beta = 0.0802$.

Although the predicted probabilities are not linearly related to x, we can transform the predicted probabilities into a new variable, z, which *is* linearly related to x. The transformation is called the *logistic* transform, and it is carried out by first finding $\hat{y}/(1-\hat{y})$ and then taking the logarithm of the result. The quantity $\hat{y}/(1-\hat{y})$ is known as the odds, and so the new variable is known as the "log-odds." Thus we have

$$z = \ln\left(\frac{\hat{y}}{1-\hat{y}}\right) = \alpha + \beta x \tag{7.15}$$

The logistic regression model is therefore one that assumes that the log-odds increases (or decreases) linearly as x increases (see Figure 7.4).

Odds are often stated in place of probabilities for events such as horse racing. If the probability that a horse wins a race is 0.2, the probability that it loses is $y=0.8$. The odds against it winning are stated as 4 to 1 (=0.8/0.2). Suppose that five people each bet $1; one bets that the horse will win, and the other four that the horse will lose. If the horse wins, the person betting for the horse will collect $5, equal to the total amount bet (of course, in reality the winner will have to give a share of his winnings to the race track, and to the government in payment of taxes!). If the probability that the horse wins rises to 0.333, the probability that it loses declines to $y = 0.667$, and the odds against it decline to 2 to 1 (=0.667/0.333). Though it is less common to do so, we could also state the odds in the other direction. When the horse has a winning probability of 0.2, the odds that the horse wins are $0.2/0.8 = 0.25$ to 1. When the probability of winning rises to 0.33, the odds in favor of the horse rise to $0.33/0.67 = 0.5$ to 1.

Returning to our example, the log-odds that an individual takes the train is given by

$$z = \ln\left(\frac{\hat{y}}{1-\hat{y}}\right) = -4.501 + 0.0802x \tag{7.16}$$

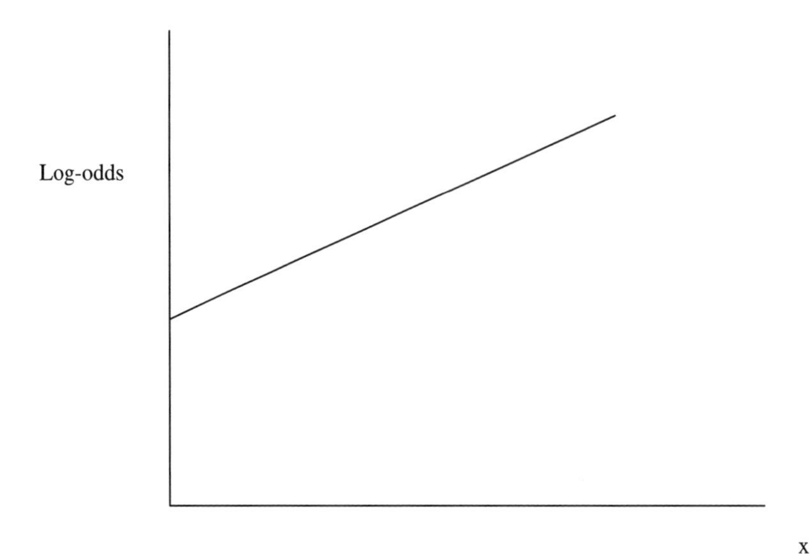

Figure 7.4 **Linear relationship between log-odds and *x***

We see that the slope coefficient β tells us how much the log-odds will change when x changes by one unit. In the present example, when $x = 32$ minutes, the predicted probability of taking the train is 0.1262, using $x = 32$ and the estimated values of α and β. The odds of taking the train are then given as 0.1444 (=0.1262/(1 – 0.1262) to 1. Stated another way, the odds against taking the train are 6.92 (=1/0.1444) to 1.

Should the individual experience an increase in auto travel time of one minute, the log of the odds of choosing the train would go up by 0.0802. What does it mean to say that the log of the odds has gone up by 0.0802? We can "undo" the log by exponentiating; if $a = \ln(b)$, then $b = e^a$. This means that if the log of the odds has increased by 0.0802, the odds of choosing the train have increased by a factor of $e^{0.0802} = 1.084$. The new odds of taking the train are now equal to $0.1565 = 0.1444 * 1.084$ to 1. Equivalently, the odds against the train have declined by a factor of 1.084, and are now 6.39 (=6.92/1.084) to 1.

When the automobile travel time increases by one minute, from $x = 32$ to $x=33$, the probability of taking the train increases from 0.1262 to 0.1353. This may be verified by using Equation 7.2 with the estimated values of α and β, and $x = 33$:

$$\hat{y}_i = \frac{e^{\alpha+\beta x_i}}{1 + e^{\alpha+\beta x_i}} = \frac{0.1565}{1 + 0.1565} = 0.1353 \tag{7.17}$$

Finally, note that in logistic regression, when $\alpha + \beta x = 0$, the predicted probability is equal to 1/2 (since $e^0 = 1$). This is equivalent to $x = -\alpha/\beta$. In our example, $-\alpha/\beta = -4.501/0.0802$, which is approximately equal to 56.

When automobile travel time is about 56 minutes, the probability that an individual takes the train is about 0.5 (or "50-50").

7.7 A Summary of Some Problems That Can Arise in Regression Analysis

Table 7.13, adapted from Haining (1990a), summarizes some of the problems that can plague regression analyses. The table describes the consequences of the problem and, in addition, describes how the problem may be diagnosed and corrected. Section numbers refer to other sections of this text that provide relevant discussion of the problem.

7.8 Multiple and Logistic Regression in *SPSS for Windows 9.0*

7.8.1 Multiple Regression

Data input is similar to that for simple linear regression (Chapter 6). Each observation is represented by a row in the data table, and each variable is represented by a column.

Click on Analyze/Regression/Linear. Then move the dependent variable into the box labeled Dependent on the right, and move the desired independent variables into the box labeled Independents. Then click on OK.

There are a number of common options that you may wish to choose before clicking on OK. Under save, it is common to check the boxes to save predicted values, residuals, leverage values to detect outliers, and confidence intervals for either the mean (i.e., the regression line) or individual predictions. All of these saved quantities will be attached as new columns in the dataset. Under statistics, it is desirable to check Collinearity diagnostics, to check for multicollinearity. Under the box where variables that are in the regression are indicated, one may choose the method by which independent variables are entered onto the right-hand side. The default is "enter", which means that all independent variables will be entered. A common alternative is to choose stepwise, which enters and removes variables one at a time, depending upon their significance.

7.8.2 Logistic Regression

There are two ways that logistic regression may be carried out using *SPSS for Windows*. The first approach is to use nonlinear least squares. This is easiest to understand, since, as the previous section indicates, we are simply looking for the values of α and β that will make the sum of squared residuals as small as possible.

Table 7.13 **Some problems that can arise in regression analysis**

Problem	Consequences	Diagnostic	Corrective action
Residuals:			
Nonnormal	Inferential tests may be invalid	Shapiro–Wilk test (4.82)	Transform *y* values
Heteroscedastic	Biased estimation of error variance, leading to invalid inference	Plot residuals against *y* and *x*s	Transform *y* values
Not independent	Underestimate-variance of regression coefficients. Inflated R^2	Moran's *I* (8.3.3)	Spatial regression (9.3)
Nonlinear relationship	Poor fit and nonindependent residuals	Scatterplots of *y* against *x*s. Added variable plots (9.2)	Transform *y* and/or *x* variables
Multicollinearity (7.1.1)	Variance of regression estimates is inflated	Variance inflation factor (7.4)	Delete variable(s)
Incorrect set of explanatory variables (7.2)	Difficulties in performing efficient analysis, and poor regression estimates	Added variable plots (9.2)	Stepwise regression (7.5)
Outliers (7.4)	May severely affect model estimates and fit	Plots. Leverage values (7.4)	Delete observations (7.4)
Categorical response variable	Linear regression model inappropriate		Logistic regression (7.6)
Spatially varying parameters geographically	Invalid estimation and inference	Moran's *I* (8.3.3)	Expansion method; Weighted regression (9.4)
Missing data at random	Could waste other case information if deleted		Estimate missing values (7.4)
Missing data (nonrandom)	Possibly invalid estimation and inference		Delete observation (7.4)

Note: Relevant sections of text are given in parentheses.
Source: Adapted from Haining (1990a), pp. 332–33.

Data input. In both cases, the approach to data input is the same. As in linear regression, the dependent variables and independent variables are arranged in columns. Each row represents an observation. It is common, but not necessary, to have the dependent variable in the first column. Make sure that the column containing the dependent variable consists of a column of "0"s and "1"s, consistent with its binary response nature.

Using SPSS for Windows 9.0 *and nonlinear least squares*

1. Choose Analyze, Regression, Nonlinear.
2. Under parameters, define α and β, and give estimated values. Choosing good estimated values is sometimes important, and not always easy to do. It may require a bit of trial and error. Using $\alpha = 0$ and $\beta = 0$ is often not a bad way to start.
3. Next, select the dependent variable.
4. Next, set up the model; this refers to the equation for the predicted values of the dependent variable. For logistic regression you should define the model as in Equation 7.13.
5. Choose OK to run the nonlinear least-squares analysis.

The nonlinear least-squares approach, however, is a bit more awkward to implement in *SPSS* than its alternative, known as the *maximum likelihood* approach to finding α and β. In addition, maximum likelihood is, to a statistician, generally a preferable alternative since it produces estimates that are unbiased and that have relatively smaller sampling variances, at least when the sample sizes are large.

Many statistical packages, including *SPSS for Windows 9.0*, make use of maximum likelihood estimation. The likelihood of observing $y = 1$ is

$$\frac{e^{\alpha+\beta x}}{1 + e^{\alpha+\beta x}}$$

Similarly, the likelihood of observing $y = 0$ is

$$1 - \frac{e^{\alpha+\beta x}}{1 + e^{\alpha+\beta x}}$$

The likelihood of the sample is therefore

$$L = \left(\frac{e^{\alpha+\beta x}}{1 + e^{\alpha+\beta x}}\right)^{\sum y_i} \left(1 - \frac{e^{\alpha+\beta x}}{1 + e^{\alpha+\beta x}}\right)^{n-\sum y_i}$$

Many programs, such as *SPSS for Windows*, choose α and β to maximize this likelihood of obtaining the observed sample.

Using logistic regression in SPSS for Windows

1. Choose Analyze, Regression, Logistic.
2. Choose the dependent variable and the covariates (independent variables).
3. Choose OK.

Table 7.14 **Logistic regresssion output**

```
Dependent Variable..    TRAIN

Beginning Block Number  0.  Initial Log Likelihood Function

-2 Log Likelihood   16.635532

* Constant is included in the model.


Beginning Block Number  1.  Method: Enter

Variable(s) Entered on Step Number
1..        AUTOTT

Estimation terminated at iteration number 4 because
Log Likelihood decreased by less than .01 percent.

 -2 Log Likelihood          12.543
 Goodness of Fit            11.629
 Cox & Snell - R^2            .289
 Nagelkerke - R^2             .385

                       Chi-Square    df Significance

 Model                      4.093    1          .0431
 Block                      4.093    1          .0431
 Step                       4.093    1          .0431

Classification Table for TRAIN
The Cut Value is .50
                        Predicted
                       .00     1.00      Percent Correct
                        0   |   1
Observed           +--------+--------+
   .00       0     |   5    |   1    |    83.33%
                   +--------+--------+
   1.00      1     |   2    |   4    |    66.67%
                   +--------+--------+
                              Overall    75.00%

---------------------- Variables in the Equation -----------------------

Variable           B        S.E.      Wald     df       Sig        R     Exp(B)

AUTOTT          .0773      .0456    2.8744      1     .0900    .2293    1.0804
Constant      -4.5362     2.7641    2.6933      1     .1008
```

Using the *SPSS for Windows* logistic regression routine, we find

$$z = -4.5362 + 0.0773x$$

Note that these values of $\alpha = -4.5362$ and $\beta = 0.0773$ are similar to those found via nonlinear least squares.

Interpreting output from logistic regression. Tables 7.14 and 7.15 display the output from the logistic regression analysis of the commuting behavior data in Table 7.12. We are, of course, interested in the slope and intercept, and these are displayed in the same part of the output where we found them in linear regression. They are given along with their standard deviations (also known as standard errors; see the column headed "S.E."). If the coefficients are more than twice their corresponding standard errors (approximately), they may be regarded as significantly different from zero. In this example, the coefficients are not significantly different from zero; this is also reflected in the column headed "Sig," where we find that the p-value associated with each coefficient is greater than 0.05.

Note that the output also contains a column headed Exp(B); this is the exponentiated slope referred to in the text, and it tells us by how much the odds will change when the x variable is increased by one unit. In this example, an increase of one minute in the commuting time leads to the odds of taking the train increasing by a factor of 1.0804.

Another interesting part of the output is the two-by-two classification table. It shows us that there were six observations where $y=0$ (the individual did not take the train). Of these, five were predicted correctly by the logistic regression equation, and one was predicted incorrectly. (A prediction is classified as "correct" if the model predicts that the actual outcome has a likelihood of

Table 7.15 **Summary of results**

		Predicted probabilities	
Y	*X*: Auto travel time (min)	Linear OLS	Logistic
0	32	.093	.113
1	89	.963	.913
0	50	.368	.339
1	49	.352	.321
0	80	.826	.839
1	56	.459	.449
0	40	.215	.191
1	70	.673	.706
1	72	.704	.737
1	76	.765	.793
0	32	.093	.113
0	58	.490	.487

greater than 0.5.) Note that the fifth individual has an observed auto travel time of $x = 80$ minutes. The model predicted that there would be a 0.839 probability that the individual would take the train (Table 7.14), yet we observed that he/she did not ($y = 0$).

Of the six individuals who *did* take the train, the model predicted four correctly, and there were two cases where the model predicted that the individual would not take the train when in fact they did. Individuals 4 and 6 both took the train, yet the model predicted probabilities of less than 0.5 that they would do so. This table summarizes how successful the model is in predicting actual outcomes.

Exercises

1. The following data are collected in a study of park attendance:

Park visit? 1 = yes; 0 = no	Distance from park (km)
0	8
0	6
1	1
0	4
1	3
0	2
0	6
1	5
1	7
1	2
1	1
1	3
1	5
1	7
0	8
0	9
0	8
0	6
1	4
1	4
0	7
0	9

Use logistic regression to determine how the likelihood of visiting the park varies with the distance that an individual resides from the park.

2. In American football, the likelihood of a successful field goal declines with increasing distance. The following data were collected one week from games played by teams in the National Football League.

Made? 1 = yes; 0 = no	Yards
0	34
1	20
0	51
1	32
0	51
0	29
1	19
0	37
0	43
1	47
1	24
1	31
1	41
1	22
1	26
1	34
1	41
1	24
1	39
1	43

(a) Use logistic regression to determine how the odds of making a field goal change as distance increases.
(b) Use the results to draw a graph depicting how the predicted probability of making a field goal changes with distance.
(c) In the waning seconds of SuperBowl XXV, Scott Norwood missed a 47-yard field goal that would have carried the Buffalo Bills to victory over the New York Giants. Use your model to predict the likelihood that a kicker is successful at a 47-yard field goal attempt. Did Norwood really deserve the criticism he received for missing the attempt?

3. What is multicollinearity? How can it be detected? Why is it a potential problem in regression analysis? How might its effects be ameliorated?

4. The number of times per year a person uses rapid transit is a linear function of income:

$$Y = 1.2 + 2.4X_1 + 8.4X_2 + 15.6X_3$$

where X_1, X_2, and X_3 are dummy variables for medium, high, and very high incomes, respectively (the low income category has been omitted). What is the predicted number of annual transit trips per year for each of the four income categories?

5. Given the following data,

Y	X
0	8
1	6
0	9
1	4
1	3

is β (the "slope" of the logistic curve) positive or negative? How do you know?

6. Suppose, for a given set a data, we find that a logistic regression yields $\beta = -0.43$. What is the change in odds for a unit change in x?

7. The following results were obtained from a regression of $n=14$ housing prices (in dollars) on median family income, size of house, and size of lot:

	Sum of squares	df	Mean square	*F*
Regression SS:	4234	3	—	—
Residual SS:	3487	—	—	
Total SS:	—	—		

	Coefficient (*b*)	Standard error (s_b)	VIF
Median family income	1.57	0.34	1.3
Size of house (sq. ft.)	23.4	11.2	2.9
Size of lot (sq. ft)	−9.5	7.1	11.3
Constant	40000	1000	

(a) Fill in the blanks.
(b) What is the value of R^2?
(c) What is the standard error of the estimate?
(d) Test the null hypothesis that $R^2=0$ by comparing the F-statistic from the table with its critical value.
(e) Are the coefficients in the direction you would hypothesize? If not, which coefficients are opposite in sign from what you would expect?
(f) Find the t-statistics associated with each coefficient, and test the null hypotheses that the coefficients are equal to zero. Use $\alpha=0.05$, and be sure to give the critical value of t.
(g) What do you conclude from the variance inflation factors (VIFs)? What modifications would you recommend in light of the VIFs?
(h) What is the predicted sales price of a house that is 1500 square feet, on a lot 60 ft × 100 ft, and in a neighborhood where the median family income is $40 000?

8. Choose a dependent variable and two or three independent variables. The variables chosen should be defined spatially. There should be at least 15 to 20, and preferably about 30 observations.

(a) State any null hypotheses you may have, as well as the alternative hypotheses.
(b) Graph the dependent variable (y) vs each independent (x) variable. Describe any obvious outliers.
(c) Graph the dependent variables against each other, and comment on any obvious multicollinearity.

(d) Regress y on each of the independent variables separately. Also regress y on the combined set of all independent variables. If you have three independent variables, you may wish to regress y on pairs of independent variables. Comment on the results.

(e) For the regression including only the single most significant independent variable,
   (i) find and graph the 95% confidence interval for the regression line;
   (ii) find and graph the 95% confidence interval for predictions of the y values.

9. Use the data in Table 7.7 to study how the number of native species on islands varies with the size of the island, the maximal elevation of the island, and the distance to nearby islands. There are many choices you will need to make; there is no single "correct" answer to this question. Some considerations you should think about include:

(a) What should be done about the missing elevation values that occur in some cases?
(b) Are there outliers? If so, how can they be identified?
(c) What about multicollinearity? Should variable(s) be eliminated from the analysis?

One goal you should have is to come up with a "best" equation, in the sense that variables in the equation are both significant and meaningful.

8 Spatial Patterns

LEARNING OBJECTIVES
- Finding geographic patterns in point and areal data
- Introduction to local statistics
- Application of Monte Carlo simulation tests to the statistical analysis of geographic clustering

8.1 Introduction

One assumption of regression analysis as applied to spatial data is that the residuals are not spatially autocorrelated – that is, there is no spatial pattern to the errors. Residuals that are not independent can affect estimates of the variances of the coefficients and hence make it difficult to judge their significance.

We have also seen in other, previous chapters that lack of independence among observations can affect the outcome of *t*-tests, ANOVA, and correlation, often leading one to find significant results where none in fact exist.

In addition to a desire to remove the complicating effects of spatially dependent observations, spatial analysts also seek to learn whether geographic phenomena cluster in space. Here they have a direct interest in the phenomenon and/or process itself, not an indirect one; the latter is the case when one wishes to correct a statistical analysis based upon spatial data. For example, crime analysts wish to know if clusters of criminal activity exist. Health officials seek to learn about disease clusters and their determinants.

In this chapter we will investigate statistical methods aimed at detecting spatial patterns and assessing their significance. The structure of the chapter follows from the fact that data are typically in the form of either point locations (where exact locations of, e.g., disease or crime are available) or in the form of aggregated areal information (where, e.g., information is available only on regional rates).

8.2 The Analysis of Point Patterns

Carry out the following experiment:

Draw a rectangle that is six inches by five inches on a sheet of paper. Locate 30 dots at random within the rectangle. This means that each dot should be located independently of the other dots. Also, for each point you locate, every subregion of a given size should have an equal likelihood of receiving the dot.

Then draw a six-by-five grid of 30 square cells on top of your rectangle. You can do this by making little tick marks at one-inch intervals along the sides of your rectangle. Connecting the tick marks will divide your original rectangle into 30 squares, each having a side of length one inch.

Give your results a score, as follows. Each cell containing no dots receives 1 point. Each cell containing one dot receives 0 points. Each cell containing two dots receives 1 point. Cells containing three dots receive 4 points, cells containing four dots receive 9 points, cells containing 5 dots receive 16 points, cells containing 6 dots receive 25 points, and cells containing 7 dots receive 36 points. Find your total score by adding up the points you have received in all thirty cells.

DO NOT READ ON UNTIL YOU HAVE COMPLETED THE INSTRUCTIONS ABOVE!

Classify your pattern as follows:

If your score is 16 or less, your pattern is significantly more uniform or regular than random.

If your score is between 17 and 45, your pattern is characterized as random.

If your score is greater than 45, your pattern exhibits significant clustering.

On average, a set of 30 randomly placed points will receive a score of 29. 95% of the time, a set of randomly placed points will receive a score between 17 and 45. The majority of people who try this experiment produce patterns that are more uniform or regular than random, and hence their scores are less than 29. Their point patterns are more spread out than a truly random pattern. When individuals see an empty space on their diagram, there is an almost overwhelming urge to fill it in by placing a dot there! Consequently, the locations of dots placed on a map by individuals are not independent of the locations of previous dots, and hence an assumption of randomness is violated.

Consider next Figures 8.1 and 8.2, and suppose you are a crime analyst looking at the spatial distribution of recent crimes. Make a photocopy of the page, and indicate in pencil where you think the clusters of crime are. Do this by simply encircling the clusters (you may define more than one on each diagram).

DO NOT READ THE NEXT PARAGRAPH UNTIL YOU HAVE COMPLETED THIS EXERCISE!

How many clusters did you find? It turns out that both diagrams were generated by locating points at random within the square! In addition to having trouble drawing random patterns, individuals also have a tendency to

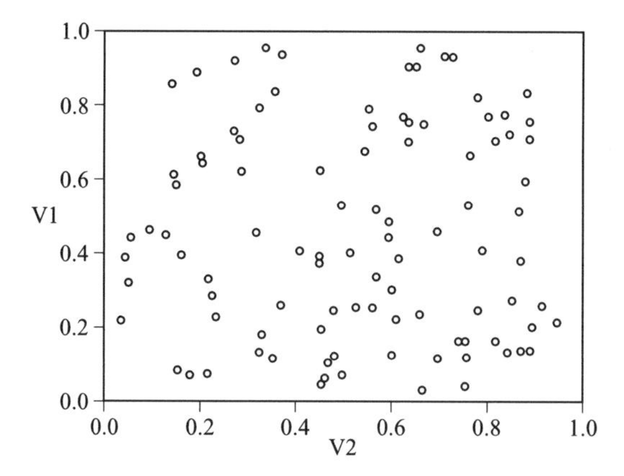

Figure 8.1 **Spatial pattern of crime**

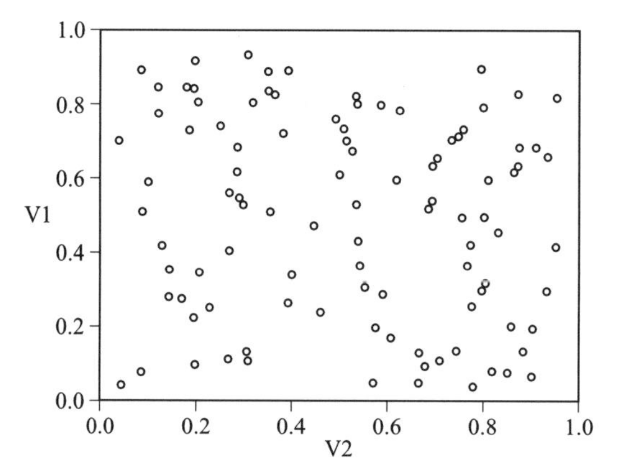

Figure 8.2 **Spatial pattern of crime**

"see" clusters where none exist. This results from the mind's strong desire to organize spatial information.

Both of these exercises point to the need for objective, quantitative measures of spatial pattern – it is simply not sufficient to rely on one's visual interpretation of a map. Crime analysts cannot necessarily pick out true clusters of crime just by looking at a map, nor can health officials always pick out significant clusters of disease from map inspection.

8.2.1 Quadrat Analysis

The experiment involving the scoring of points falling within the "6 × 5" rectangle is an example of *quadrat analysis*, developed primarily by ecologists in the first half of the twentieth century. In quadrat analysis, a grid of square cells of equal size is used as an overlay on top of a map of incidents. One then counts the number of incidents in each cell. In a random pattern, the mean number of points per cell will be roughly equal to the variance of the number of points per cell.

If there is a large amount of variability in the number of points from cell to cell (some cells have many points, some have none, etc.), this implies a tendency toward clustering. If there is very little variability in the number of points from cell to cell, this implies a tendency toward a systematic pattern (where the number of points per cell would be the same). The statistical test makes use of a chi-square statistic involving the variance–mean ratio:

$$\chi^2 = \frac{(m-1)s^2}{\bar{x}} \tag{8.1}$$

where m is the number of quadrats, and $\bar{x}$ and s^2 are the mean and variance of the number of points per quadrat, respectively. This value is then compared with a critical value from a chi-square table with $m - 1$ degrees of freedom.

Quadrat analysis is easy to employ, and it has been a mainstay in the spatial analyst's toolkit of pattern detectors over several decades. One important issue is the size of the quadrat; if the cell size is too small, there will be many empty cells, and if clustering exists on all but the smallest spatial scales it will be missed. If the cell size is too large, one may miss patterns that occur *within* cells. One may find patterns on some spatial scales and not at others, and thus the choice of quadrat size can seriously influence the results. Curtiss and McIntosh (1950) suggest an "optimal" quadrat size of two points per quadrat. Bailey and Gatrell (1995) suggest that the mean number of points per quadrat should be about 1.6.

Summary of the quadrat method

(1) Divide a study region into m cells of equal size.
(2) Find the mean number of points per cell ($\bar{x}$). This is equal to the total number of points divided by the number of cells (m).
(3) Find the variance of the number of points per cell, s^2, as follows:

$$s^2 = \frac{\sum_{i=1}^{i=m} (x_i - \bar{x})^2}{m-1} \tag{8.2}$$

where x_i is the number of points in cell i.
(4) Calculate the variance–mean ratio (VMR):

$$\text{VMR} = \frac{s^2}{\bar{x}} \tag{8.3}$$

(5) Interpret the results as follows.

If $s^2/\bar{x} < 1$, the variance of the number of points is less than the mean. In the extreme case where the ratio approaches zero, there is very little variation in the number of points from cell to cell. This characterizes situations where the distribution of points is spread out, or uniform, across the study area.

If $s^2/\bar{x} > 1$, there is a good deal of variation in the number of points per cell – some cells have substantially more points than expected (i.e., $x_i > \bar{x}$ for some cells i), and some cells have substantially fewer than expected (i.e., $x_i < \bar{x}$). This characterizes situations where the point pattern is more clustered than random. A value of $s^2/\bar{x}$ near one indicates that the points are close to being randomly distributed across the study area.

Hypothesis Testing. How can we be more precise in testing the null hypothesis that there is no spatial pattern? Suppose we were to simulate the null hypothesis by placing points at random in a study area, and that we then carried out the procedure described above for finding the variance–mean ratio. Furthermore, suppose we were to repeat this many times (say 1000), and then draw a histogram of the results. We would find that the mean of our 1000 VMR values would be near one, and that the histogram would be asymmetric, displaying a positive skew (see Figure 8.3). Values of VMR in the tails of the histogram (also known as the *sampling distribution* of VMR), indicate values that are relatively rare when the underlying null hypothesis of no pattern is true.

For an actual set of observed data, we decide to accept the null hypothesis that the points are randomly distributed in space if the VMR for the observed data does not differ too much from one; otherwise, we reject the null hypothesis. More specifically, if the VMR for an observed pattern is greater than VMR_H (shown in Figure 8.3), the null hypothesis is rejected, and the pattern is taken to be more uniform than random. Similarly, if the observed VMR is less than VMR_L, the null hypothesis is rejected, and the pattern is taken to be more clustered than random.

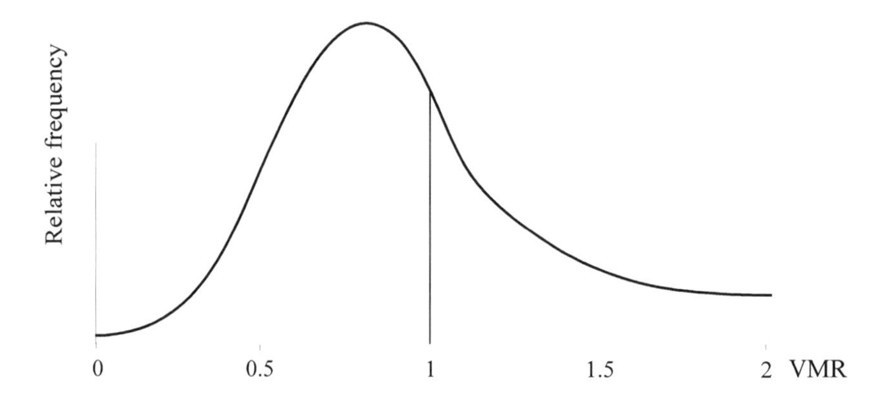

Figure 8.3 **Sampling distribution of VMR when H_0 is rue**

If we were to actually observe an extreme value of VMR in our data (either greater than VMR_H or less than VMR_L), we reject the null hypothesis that the pattern is random. In this case, either (a) the null hypothesis is actually true (in which case we have incorrectly rejected it, and committed a Type I error), or (b) the null hypothesis is not true, and we have made a correct decision. To establish the critical, cutoff values, VMR_L and VMR_H, we first have to decide upon how great a likelihood of a Type I error that we are willing to tolerate.

If we use $\alpha = 0.05$, then the 50 most extreme values out of the total of 1000 in our experiment are used to obtain the critical values (since $50/1000 = 0.05$). If we rank the 1000 VMR values from lowest to highest, the 25th VMR on our list would be chosen as VMR_L; 25 out of 1000 times we will observe a lower VMR than this when H_0 is true. Similarly, the 975th VMR on our ordered list would be chosen as VMR_H; 25 out of 1000 times we can expect to observe a VMR higher than this when H_0 is true. Thus 50 out of 1000, or 5% of the time we will incorrectly reject a true hypothesis when we use these critical values. In those 50 instances, we would make a Type I error, since we would reject H_0 when in fact it was true, and we had simply observed an unusual value of VMR in the tail of the sampling distribution.

Example. We wish to know whether the pattern observed in Figure 8.4 is consistent with the null hypothesis that the points were placed at random. We first calculate the VMR. There are 100 points on the 10×10 grid, implying a mean of one point per cell. There are 6 cells with 3 points, 20 cells with 2 points, 42 cells with one point, and 32 cells with no points. The variance is

$$\left\{6(3-1)^2 + 20(2-1)^2 + 42(1-1)^2 + 32(0-1)^2\right\}/99 = 76/99 = 0.77 \quad (8.4)$$

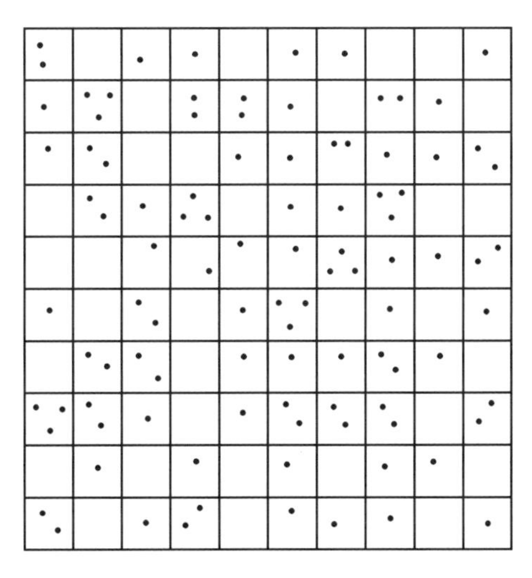

Figure 8.4 **A spatial point pattern**

and, since the mean is equal to one, this is also our observed VMR. Since VMR < 1, there is a tendency toward a uniform pattern. How unlikely is a value 0.77 if the null hypothesis is true – is it unlikely enough that we should reject the null hypothesis?

To assess this, we need to find the sampling distribution of VMR, when H_0 is true. One hundred points were assigned to cells at random in a 10×10 grid. The VMR was calculated using $\bar{x} = 1$, since there is an average of one point per cell. This was repeated 1000 times to establish the form of the sampling distribution, when the null hypothesis of a random point pattern is true. The resulting 1000 VMRs were ranked from lowest to highest. The 25th lowest value was $\text{VMR}_\text{L} = 0.747$ and the 975th value on the list was $\text{VMR}_\text{H} = 1.313$.

These critical values can then be used to decide whether the actual pattern of interest exhibits significant deviations from randomness. Since our observed value of VMR = 0.77 is not less than the lower critical value of 0.7475, we accept the null hypothesis: a VMR of 0.77 is not particularly unusual when H_0 is true.

The process of deriving a sampling distribution via simulation of the null hypothesis in the manner we have just described is known as the *Monte Carlo* method. It has the merit of making the underlying ideas associated with hypothesis testing easy to convey and, hopefully, easy to understand. But if we had ten people each using the Monte Carlo method to find critical values of VMR in this example, we would get ten different sets of critical valucs. Thc larger the number of repetitions, the closer together the sets of critical values would be. Another feature of the Monte Carlo method in this example is that the critical values we found are only appropriate when there are 100 points in 100 cells. If we were to look at another example with other than 100 cells, we would need to repeat a large number of simulations under the null hypothesis to establish critical values.

There is an easy way to avoid having to actually simulate the sampling distribution under the assumption that H_0 is true. The critical values can be determined by using the fact that the quantity $\chi^2 = (m - 1)\text{VMR}$ has a chi-square distribution, with $m - 1$ degrees of freedom, when H_0 is true. This fact allows us to obtain critical values, χ^2_L and χ^2_H, from a chi-square table. In particular, we will reject H_0 if either $\chi^2 < \chi^2_\text{L}$ or $\chi^2 > \chi^2_\text{H}$.

When the number of degrees of freedom (df) is large, the sampling distribution of $\chi^2 = (m - 1)\text{VMR}$ begins to approach the shape of a normal distribution. In particular, when df > 30 or so $(m - 1)\text{VMR}$ will, when H_0 is true, have a normal distribution with mean $m - 1$ and variance equal to $2(m - 1)$. This means that we can treat the quantity

$$z = \frac{(m-1)\text{VMR} - (m-1)}{\sqrt{2(m-1)}} = \sqrt{(m-1)/2}\,(\text{VMR} - 1) \qquad (8.5)$$

as a normal random variable with mean 0 and variance 1. With $\alpha = 0.05$, the critical values are $z_\text{L} = -1.96$ and $z_\text{H} = +1.96$. The null hypothesis of no pattern is rejected if $z < z_L$ (implying clustering) or if $z > z_H$ (implying uniformity). In our example, we have

$$z = \frac{99(0.77) - 99}{\sqrt{2(99)}} = \sqrt{99/2}(0.77 - 1) = -1.618 \tag{8.6}$$

This also falls within the critical values of z and hence we do not have strong enough evidence to reject the null hypothesis.

If cells of a different size had been used, the results, and possibly the conclusions, would have been different. By aggregating the cells in Figure 8.4 to a 5×5 grid of 25 cells, the VMR declines to 0.6875 (based on a variance of 1.658^2 and a mean of 4 points per cell). The χ^2 value is $24(0.6875) = 16.5$. Since the number of degrees of freedom is less than 30, we will use the chi-square table (Table A.5) to assess significance. With 24 degrees of freedom, and using interpolation to find the critical values at $p = 0.025$ and $p = 0.975$, yields $\chi^2_L = 12.0$ and $\chi^2_H = 40.5$. Since our observed value falls between these limits, we again fail to reject the hypothesis of randomness.

To summarize, after finding VMR, steps 1–4 above, calculate $\chi^2 = (m - 1)\text{VMR}$, and compare it with the critical values found in a chi-square table using $\text{df} = m - 1$. If $m - 1$ is greater than about 30, you can use the fact that $z = \sqrt{(m - 1)/2}(\text{VMR} - 1)$ has a normal distribution with mean 0 and variance 1, implying that, for $\alpha = 0.05$, one can compare z with the critical values $z_L = -1.96$ and $z_H = +1.96$.

It is interesting to note that the quantity $\chi^2 = (m - 1)\text{VMR}$ may be written as

$$\chi^2 = (m - 1)\text{VMR} = \frac{(m - 1)s^2}{\bar{x}} = \frac{(m - 1)\sum(x_i - \bar{x})^2}{\bar{x}(m - 1)} = \frac{\sum(x_i - \bar{x})^2}{\bar{x}} \tag{8.7}$$

The quantity $\sum(x_i - \bar{x})^2/\bar{x}$ is the sum across cells of the squared deviations of the observed numbers from the expected numbers of points in a cell, divided by the expected number of points in a cell. This is commonly known as the chi-square goodness-of-fit test.

8.2.2 Nearest Neighbor Analysis

Clark and Evans (1954) developed nearest neighbor analysis to analyze the spatial distribution of plant species. They developed a method for comparing the observed average distance between points and their nearest neighbors with the distance that would be expected between nearest neighbors in a random pattern. The nearest neighbor statistic, R, is defined as the ratio between the observed and expected values:

$$R = \frac{R_0}{R_e} = \frac{\bar{x}}{1/(2\sqrt{\lambda})} \tag{8.8}$$

where $\bar{x}$ is the mean of the distances of points from their nearest neighbors and λ is the number of points per unit area. R varies from 0 (a value obtained when all points are in one location, and the distance from each point to its nearest neighbor is zero) to a theoretical maximum of about 2.14 (for a perfectly uniform or systematic pattern of points spread out on an infinitely large two-dimensional plane). A value of $R = 1$ indicates a random pattern, since the observed mean distance between neighbors is equal to that expected in a random pattern. It is also known that if we examined many random patterns, we would find that the variance of the mean distances between nearest neighbors is

$$V[R_e] = \frac{4 - \pi}{4\pi\lambda n} \tag{8.9}$$

where n is the number of points. Thus we can form a z-test to test the null hypothesis that the pattern is random:

$$z = \frac{(R_0 - R_e)}{\sqrt{V[R_e]}} = \frac{R_0 - R_e}{\sqrt{(4 - \pi)/(4\pi\lambda n)}} = 3.826(R_0 - R_e)\sqrt{\lambda n} \tag{8.10}$$

The quantity z has a normal distribution with mean 0 and variance 1, and hence tables of the standard normal distribution may be used to assess significance. A value of $z > 1.96$ implies that the pattern has significant uniformity, and a value of $z < -1.96$ implies that there is a significant tendency toward clustering.

The strength of this approach lies in its ease of calculation and comprehension. Several cautions should be noted in the interpretation of the nearest neighbor statistic. The statistic, and its associated test of significance, may be affected by the shape of the region. Long, narrow, rectangular shapes may have relatively low values of R simply because of the constraints imposed by the region's shape. Points in long, narrow rectangles are *necessarily* close to one another. Boundaries can also make a difference in the analysis. One solution to the boundary problem is to place a buffer area around the study area. Points inside the study area may have nearest neighbors that fall within the buffer area, and these distances (rather than distances to those points that are nearest within the study area) should be used in the analysis.

Another potential difficulty with the statistic is that, since only nearest neighbor distances are used, clustering is only detected on a relatively small spatial scale. To overcome this it is possible to extend the approach to second- and higher-order nearest neighbors.

It is often of interest to ask not only whether clustering exists, but whether clustering exists over and above some background factor (such as population). Nearest neighbor methods are not particularly useful in these situations because they only relate to spatial location and not to other attributes.

The approaches to the study of pattern that are described in the next section do not have this limitation.

Illustration. For the point pattern in Figure 8.5, distances are given along the lines connecting the points. The mean distance between nearest neighbors is $R_0 = (1+2+3+1+3+3)/6 = 13/6 = 2.167$. The expected mean distance between nearest neighbors in a pattern of six points placed randomly in a study region with area $7 \times 6 = 42$ is

$$R_e = 1/(2\sqrt{\lambda}) = 1/(2\sqrt{6/42}) = 1.323 \quad (8.11)$$

The nearest neighbor statistic is $R = 2.167/1.323 = 1.638$, which means that the pattern displays a tendency toward uniformity. To assess significance, we can calculate the z-statistic from (8.10) as $3.826\ (2.167 - 1.323)\ \sqrt{6/42 * 6} = 2.99$, which is much greater than the critical value of 1.96; this implies rejection of the null hypothesis of a random pattern. However, we have neglected boundary effects, and these have a significant effect. As an alternative way to test the null hypothesis, we can randomly choose 6 points by choosing random x-coordinates in the range (0,7) and random y-coordinates in the range (0,6). Then we compute the mean distance to nearest neighbor, and repeat the whole process many times. Simulating the random placement of 6 points in the 7×6 study region 10 000 times led to a mean distance between nearest neighbors of 1.62. This is greater than the expected distance of $R_e = 1.323$ noted above. This greater-than-expected distance can be attributed directly to the fact that points near the border of the study region are relatively farther from other points in the study region than they presumably would have been to points just outside of the study region. Ordering the 10 000 mean distances to nearest neighbors

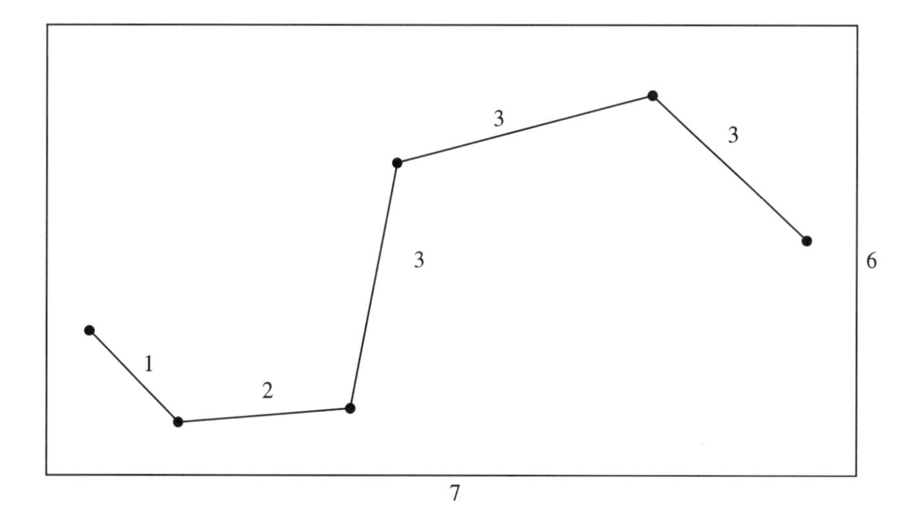

Figure 8.5 **Nearest neighbor distances**

reveals that the 9500th highest one is 2.29. Only 5% of the time would we expect a mean distance greater than 2.29. Our observed distance of 2.167 is less than 2.29, and so we, having accounted for boundary effects through our Monte Carlo simulation, accept the null hypothesis.

8.3 Geographic Patterns in Areal Data

8.3.1 An Example Using a Chi-Square Test

In a regression of housing prices on housing characteristics, suppose that we have 51 observations that have been categorized into three spatial locations (neighborhoods). How might we tell whether there is a tendency for positive or negative residuals to cluster in one or more neighborhoods? One idea is to note whether each residual is positive or negative, and then to tabulate the residuals by neighborhood (see the hypothetical data in Table 8.1).

We can use a chi-square test to determine whether there is any neighborhood-specific tendency for residuals to be positive or negative. Under the null hypothesis of no pattern, the expected values are equal to the product of the row and column totals, divided by the overall total. These expected values are given in parentheses in Table 8.2.

The chi-square statistic is

$$\chi^2 = \sum_{i=1}^{n} \frac{(O - E)^2}{E} \tag{8.12}$$

where O is the observed frequency and E is the expected frequency. In this example, the value of chi-square is 4.40 (see inset), which is less than the critical value of 5.99, using $\alpha = 0.05$ and 2 degrees of freedom (the number of degrees

Table 8.1 **Hypothetical residuals**

	Neighborhood			
	1	2	3	Total
+	10	6	7	23
–	6	15	7	28
Total	16	21	14	51

Table 8.2 **Observed and expected frequencies of residuals**

	Neighborhood			
	1	2	3	Total
+	10	6	7	23
	(7.22)	(9.47)	(6.31)	
–	6	15	7	28
	(8.78)	(11.53)	(7.69)	
Total	16	21	14	51

of freedom is equal to the number of rows minus one, times the number of columns minus one). Therefore the null hypothesis of no pattern is not rejected.

INSET: The observed chi-square statistic for the data in Table 8.2:

$$\chi^2 = \frac{(10-7.22)^2}{7.22} + \frac{(6-9.47)^2}{9.47} + \frac{(7-6.31)^2}{6.31} + \frac{(6-8.78)^2}{8.78} + \frac{(15-11.53)^2}{11.53} + \frac{(7-7.69)^2}{7.69} = 4.40$$

When spatial autocorrelation is detected, what can be done about it? One idea is to include a new, location-specific dummy variable. This will serve to capture the importance of an observation's location in a particular neighborhood. In our housing price example, we could add two variables, one for two of the three neighborhoods (following the usual practice of omitting one category). You should also note that if there are k neighborhoods, it is *not* necessary to have $k-1$ dummy variables; rather, you might choose to have only one or two dummy variables for those neighborhoods having large deviations between the observed and predicted values.

8.3.2 The Join-Count Statistic

Whether areas of positive or negative residuals cluster on the map can be determined by first asking how many total "joins" there are (i.e., the total number of cases where two subareas share a common boundary). Each join is then classified as a "++", "+−", or "−−", depending upon the signs of the residuals in the two areas. For example, the five-zone system in Figure 8.6 has three zones with negative residuals and two with positive residuals. There is a total of seven joins (i.e., pairs of regions that share a common boundary). One of these joins is a "++" join, one is a "−−" join, and the remaining five joins are "+−" joins.

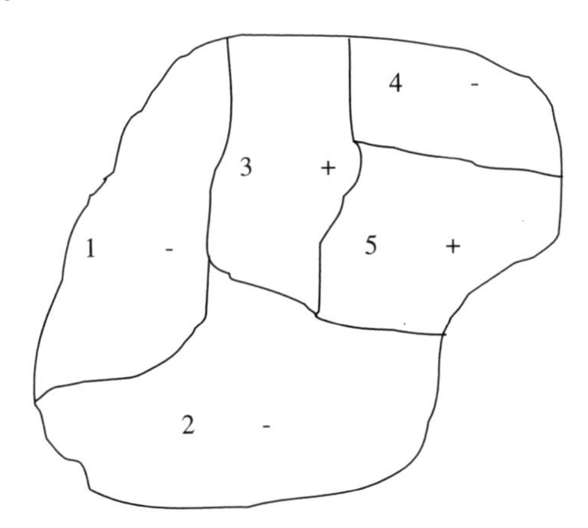

Figure 8.6 **Positive and negative residuals in a five-region system**

The join count statistic compares the observed number of +− joins with the number of +− joins that would be expected if no spatial autocorrelation were present. The expected number of +− joins is

$$E[+-] = \frac{2JPM}{N(N-1)} \tag{8.13}$$

where J is the total number of joins, P is the number of positive residuals, M is the number of negative ("minus") residuals, and N is the total number of areas ($N = P + M$). For the system in Figure 8.6,

$$E[+-] = \frac{2(7)(2)(3)}{5(4)} = 4.2 \tag{8.14}$$

The variance of the number of "+−" joins is equal to the complex expression

$$\begin{aligned} V[+-] =& E[+-] - E[+-]^2 + \frac{\sum_i L_i(L_i - 1)PM}{N(N-1)} \\ &+ \frac{4[J_i(J_i - 1) - \sum_i L_i(L_i - 1)]P(P-1)M(M-1)}{N(N-1)(N-2)(N-3)} \end{aligned} \tag{8.15}$$

where L_i is the number of links (joins) from region i to other regions. For Figure 8.6, the values of L_i are 2, 3, 4, 2, and 3, for $i = 1$, 2, 3, 4, and 5, respectively. Using Equation 8.15, for the zonal system in Figure 8.6 we have

$$\mathrm{V}[+-] = 4.2 - 4.2^2 + \frac{28(2)(3)}{5(4)} + \frac{4[(7)(6) - 28](2)(1)(3)(2)}{5(4)(3)(2)} = 0.56 \tag{8.16}$$

The z-statistic

$$z = \frac{(\text{Obs. } ``+-") - E[+-]}{\sqrt{\mathrm{V}[+-]}} \tag{8.17}$$

has a normal distribution with mean zero and variance one. Thus tables of the normal distribution can be used to test the null hypothesis that the spatial pattern is random. For our example, we have slightly more +− joins than expected. Clearly there is no clustering of positive or negative residuals on the map, but might the "checkerboard" pattern characterized by "+"s being next to "−"s be significant? The z-statistic is

$$z = \frac{5 - 4.2}{\sqrt{.56}} = 1.07 \tag{8.18}$$

which is less than the critical value, indicating that the null hypothesis of randomly placed residuals cannot be rejected.

8.3.3 Moran's I

The join count statistic, if used to evaluate the presence or absence of spatial autocorrelation, has the drawback of not using all of the available information – that is, it makes use only of the signs of the residuals and not their magnitude. Moran's I statistic is an alternative measure of spatial autocorrelation.

Moran's I statistic (1948, 1950) is one of the classic (as well as one of the most common) ways of measuring the degree of spatial autocorrelation in areal data. Moran's I is calculated as follows:

$$I = \frac{n \sum_{i}^{n} \sum_{j}^{n} w_{ij}(y_i - \bar{y})(y_j - \bar{y})}{\left(\sum_{i}^{n} \sum_{j}^{n} w_{ij}\right) \sum_{i}^{n} (y_i - \bar{y})^2} \tag{8.19}$$

where there are n regions and w_{ij} is a measure of the spatial proximity between regions i and j. It is interpreted much like a correlation coefficient. Values near $+1$ indicate a strong spatial pattern (high values tend to be located near one another, and low values tend to be located near one another). Values near -1 indicate strong negative spatial autocorrelation; high values tend to be located near low values. (Spatial patterns with negative autocorrelation are either extremely rare or nonexistent!) Finally, values near 0 indicate an absence of spatial pattern.

Though perhaps daunting at first glance, it is helpful to realize that if the variable of interest is first transformed into a z-score $\{z = (x - \bar{x})/s\}$, a much simpler expression for I results:

$$I = \frac{n \sum_{i} \sum_{j} w_{ij} z_i z_j}{(n-1) \sum_{i} \sum_{j} w_{ij}} \tag{8.20}$$

The conceptually important part of the formula is the numerator, which sums the products of z-scores in nearby regions. Pairs of regions where *both* regions exhibit above-average scores (or below average scores) will contribute positive terms to the numerator, and these pairs will therefore contribute toward positive spatial autocorrelation. Pairs where one region is above average and the other is below average will contribute negatively to the numerator, and hence to negative spatial autocorrelation.

The weights $\{w_{ij}\}$ can be defined in a number of ways. Perhaps the most common definition is one of *binary connectivity*; $w_{ij}=1$ if regions i and j are contiguous, and $w_{ij}=0$ otherwise. Sometimes the w_{ij} defined in this way are standardized to define new w_{ij}^* by dividing by the number of regions i is connected to; i.e., $w_{ij}^* = w_{ij}/\sum_j w_{ij}$. In this case all regions i are characterized by a set of weights linking i to other regions that sum to one; i.e., $\sum_j w_{ij}^* = 1$.

Alternatively, $\{w_{ij}\}$ may be defined as a function of the distance between i and j (e.g., $w_{ij} = d_{ij}^{-\beta}$ or $w_{ij} = \exp[-\beta d_{ij}]$), where the distance between i and j

could be measured along the line connecting the centroids of the two regions. It is conventional to use $w_{ij} = 0$. It is also common, though not necessary, to use symmetric weights, so that $w_{ij} = w_{ji}$.

It is important to recognize that the value of I is very dependent upon the definition of the $\{w_{ij}\}$. Using a simple binary connectivity definition for the map in Figure 8.6 gives us

$$\mathbf{W} = \{w_{ij}\} = \begin{matrix} 0 & 1 & 1 & 0 & 0 \\ 1 & 0 & 1 & 1 & 0 \\ 1 & 1 & 0 & 1 & 1 \\ 0 & 1 & 1 & 0 & 1 \\ 0 & 0 & 1 & 1 & 0 \end{matrix}$$

In this instance, the definition of $\{w_{ij}\}$ causes the neighborhood around region 1 to be much smaller than the neighborhood around region 2 or 3. This is not necessarily "wrong", but suppose that we were interested in the spatial autocorrelation of a disease that was characterized by rates that were strongly associated over small spatial scales but not correlated over large spatial scales. If we expect disease rates in regions 1 and 2 to be highly correlated while we expect those in regions 4 and 5 to be uncorrelated due to their large spatial separation, our observed value of I will be a combined measure of strong association between close pairs and weak association between distant pairs. For this example, it might be more appropriate to use a distance-based definition of $\{w_{ij}\}$.

Illustration. Consider the six-region system in Figure 8.7. Using a binary connectivity definition of the weights leads to:

$$\mathbf{W} = \begin{bmatrix} 0 & 1 & 1 & 0 & 0 & 0 \\ 1 & 0 & 1 & 1 & 1 & 0 \\ 1 & 1 & 0 & 0 & 1 & 1 \\ 0 & 1 & 0 & 0 & 1 & 0 \\ 0 & 1 & 1 & 1 & 0 & 1 \\ 0 & 0 & 1 & 0 & 1 & 0 \end{bmatrix} \tag{8.21}$$

where an entry in row i and column j is denoted by w_{ij}. The double summation in the numerator of I (see Equation 8.19) is found by taking the product of the deviations from the mean for all pairs of adjacent regions

$$\begin{aligned} &(32-21)(26-21) + (32-21)(19-21) + (26-21)(32-21) \\ &\quad + (26-21)(19-21) + (26-21)(18-21) + (26-21)(17-21) \\ &\quad + (19-21)(32-21) + (19-21)(26-21) + (19-21)(17-21) \\ &\quad + (19-21)(14-21) + (18-21)(26-21) + (18-21)(17-21) \\ &\quad + (17-21)(19-21) + (17-21)(19-21) + (17-21)(26-21) \\ &\quad + (17-21)(18-21) + (17-21)(14-21) + (14-21)(19-21) \\ &\quad + (14-21)(17-21) = 100 \end{aligned} \tag{8.22}$$

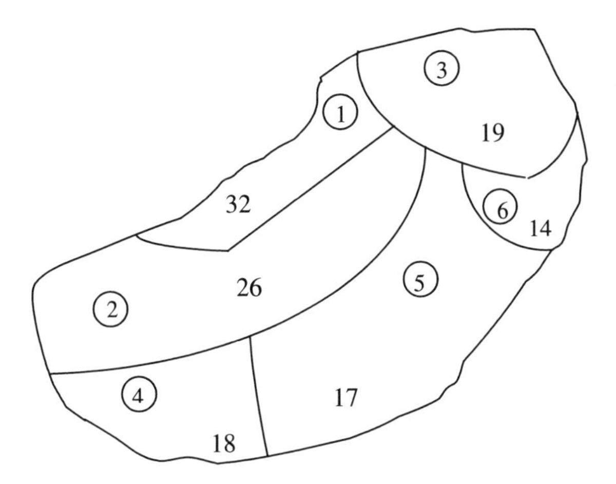

Figure 8.7 **Hypothetical six-region system**

Since the sum of the weights in (8.21) is 18, and since the variance of the regional values is 224/5, Moran's I is equal to

$$I = \frac{6(100)}{18(224)} = 0.1488 \tag{8.23}$$

In addition to this descriptive interpretation, there is a statistical framework that allows one to decide whether any given pattern deviates significantly from a random pattern. If the number of regions is large, the sampling distribution of I, under the hypothesis of no spatial pattern, approaches a normal distribution, and the mean and variance of I can be used to create a Z-statistic in the usual way:

$$Z = \frac{I - E[I]}{\sqrt{V[I]}} \tag{8.24}$$

The value is then compared with the critical value found in the normal table (e.g., $\alpha = 0.05$ would imply critical values of -1.96 and $+1.96$). The mean and variance are equal to

$$\left.\begin{aligned} E[I] &= \frac{-1}{n-1} \\ V[I] &= \frac{n^2(n-1)S_1 - n(n-1)S_2 + 2(n-2)S_0^2}{(n+1)(n-1)^2 S_0} \end{aligned}\right\} \tag{8.25}$$

where

$$\left.\begin{aligned} S_0 &= \sum_i^n \sum_{j \neq i}^n w_{ij} \\ S_1 &= 0.5 \sum_i^n \sum_{j \neq i}^n (w_{ij} + w_{ji})^2 \\ S_2 &= \sum_k^n \left(\sum_j^n w_{kj} + \sum_i^n w_{ik} \right)^2 \end{aligned}\right\} \qquad (8.26)$$

Computation is not complicated, but it is tedious enough for one to not want to do it by hand! Unfortunately, few software packages that calculate the coefficient and its significance are available. Exceptions include Anselin's (1992) *SpaceStat* and the CrimeStat package downloadable from www.icpsr.umich.edu/NACJD/crimestat.html

Fortunately, there are also simplifications and approximations that facilitate the use of Moran's I. An alternative way of finding Moran's I is to simply take the ratio of two regression slope coefficients (see Griffith 1996). The numerator of I is equal to the regression slope obtained when the quantity $a_i = \sum_{j=1}^n w_{ij} z_j$ is regressed on z_i, and the denominator of I is equal to the regression slope obtained when the quantity $b_i = \sum_{j=1}^n w_{ij}$ is regressed on $c_i = 1$. The zs represent the z-scores of the original variables, and the slope coefficients are found using no-intercept regression (i.e., constraining the result of the regression so that the intercept is equal to zero).

In addition, Griffith gives $2/\sum\sum w_{ij}$ as an approximation for the variance of the Moran coefficient. This expression, though it works best only when the number of regions is sufficiently large (about 20 or more), is clearly easier to compute than the alternative given in Equations 8.25 and 8.26! If observational units are on a square grid and connectivity is indicated by the four adjacent cells, the variance may be approximated by $1/(2n)$, where n is the number of cells. Based on either a grid of hexagonal cells or a map displaying "average" connectivity with other regions, the variance may be approximated by $1/(3n)$. An example is given in Section 8.5.

The use of the normal distribution to test the null hypothesis of randomness relies upon one of two assumptions:

(1) *Normality*. It can be assumed that regional values are generated from identically distributed normal random variables (i.e., the variables in each region arise from normal distributions that have the same mean and same variance in each region).
(2) *Randomization*. It can be assumed that all possible permutations (i.e., regional rearrangements) of the regional values are equally likely.

The formulae given above (Equations 8.25 and 8.26) for the variance assumes that the normality assumption holds. The variance formula for the randomization assumption is algebraically more complex, and gives values that are only slightly different from that given in 8.25 and 8.26 (see, e.g., Griffith 1987).

If either of the two assumptions above holds, the sampling distribution of I will have a normal distribution if the null hypothesis of no pattern is true. One of the two assumptions must hold to generate the sampling distribution of I so that critical values of the test statistic may be established. For example, if the first assumption were used to generate regional values, I could be computed; this could then be repeated many times, and a histogram of the results could be produced. The histogram would have the shape of a normal distribution, a mean of $E[I]$, and a variance of $V[I]$. Similarly, the observed regional values on a map could be randomly rearranged many times, and the value of I computed each time. Again, a histogram could be produced; it would again have the shape of a normal distribution with mean $E[I]$ and a variance slightly different from $V[I]$. If we can rely on one of these two assumptions we do not need to perform these experiments to generate histograms, since we know beforehand that they will produce normal distributions with known mean and variance.

Unfortunately, there are many circumstances in geographical applications that lead the analyst to question the validity of either assumption. For example, maps of counties by township are often characterized by high population densities in the townships corresponding to or adjacent to the central city, and by low population densities in outlying townships. Rates of crime or disease, though they may have equal means across townships, are unlikely to have equal variances. This is because the outlying towns are characterized by greater uncertainty – they are more likely to experience atypically high or low rates simply because of the chance fluctuations associated with a relatively smaller population base. Thus assumption 1 is not satisfied, since all regional values do not come from identical distributions – some regional values, namely the outlying regions, are characterized by higher variances. Likewise, not all permutations of regional values are equally likely – permutations with atypically high or low values out in the periphery are more likely than permutations with atypically high or low values near the center.

How can we test the null hypothesis of no spatial pattern in this instance? One approach is to use Monte Carlo simulation. Suppose we have data on the number of diseased individuals (n_i) and the population (p_i) in each region. Since the Z-test described above is no longer valid, we need an alternative way to come up with critical values. The null hypothesis of no spatial pattern in disease rates can be assessed via simulation. Assign the disease to an individual with probability $\sum_i n_i / \sum_i p_i$. Then calculate Moran's I. This is repeated many times, and the resulting values of Moran's I may be used to create a histogram depicting the relative frequencies of I when the null hypothesis is true. Furthermore, the values can be arranged from lowest to highest, and this list can be used to find critical values of I. For example, if the simulations are carried out 1000 times, and critical values are desired for a test using $\alpha = 0.05$, they can found from the ordered list of I values. The lower critical value would be the 25th item on the list, and the upper critical value would be the 975th item on the list.

Illustration of the Monte Carlo method. Dominik Hasek, the goalie for the gold-medal Czech ice hockey team in the 1998 Olympics, saves 92.4% of all shots he faces when he plays professionally for the Buffalo Sabres of the National Hockey League (NHL). The average save percentage of other goalies in the NHL is 90%. Hasek tends to face about 31 shots per game, while the Sabres manage just 25 shots per game on the opposing goalie. To evaluate how much Hasek means to the Sabres, compare the outcomes of 1000 games using Hasek's statistics with the outcomes of 1000 games assuming the Sabres had an "average" goalie who stops 90% of the shots against him.

Solution. Take 31 random numbers between 0 and 1. Count those greater than 0.924 as goals against the Sabres with Hasek. Take 25 numbers from a uniform distribution between 0 and 1, and count those greater than 0.9 as goals for the Sabres. Record the outcome (win, loss, or tie). Repeat this 1000 times (preferably using a computer!), and tally the outcomes. Finally, repeat the entire experiment using random numbers greater than 0.9 (instead of 0.924) to generate goals against the Sabres without Hasek. Each time the experiment is performed, a different outcome will be obtained. In one comparison, the results were as follows:

	Wins	Losses	Ties
Scenario 1 (with Hasek)	434	378	188
Scenario 2 (without Hasek)	318	515	167

To evaluate Hasek's value to the team over the course of an 82-game season, the outcomes above may first be converted to percentages, multiplied by 82, and then rounded to integers, yielding:

	Wins	Losses	Ties
Scenario 1	36	31	15
Scenario 2	26	42	14

Thus Hasek is "worth" about 10 wins; that is, they win about ten games a year that they would have lost if they had an "average" goalie.

8.4 Local Statistics

8.4.1 Introduction

Besag and Newell (1991) classify the search for clusters into three primary areas. First are "general" tests, designed to provide a single measure of overall pattern for a map consisting of point locations. These general tests are intended to provide a test of the null hypothesis that there is no underlying pattern, or deviation from randomness, among the set of points. Examples include the nearest neighbor test, the quadrat method, and the Moran statistic, all outlined above. In other situations, the researcher wishes to know whether there is a cluster of events around a single or small number of prespecified foci. For example, we may wish to know whether disease clusters around a toxic waste site, or whether crime clusters around a set of liquor establishments. Finally, Besag and Newell describe "tests for the detection of clustering." Here there is no *a priori* idea of where the clusters may be; the methods are aimed at searching the data and uncovering the size and location of any possible clusters.

General tests are carried out with what are called "global" statistics; again, a single summary value characterizes any deviation from a random pattern. "Local" statistics are used to evaluate whether clustering occurs around particular points, and hence are employed for both focused tests and tests for the detection of clustering. Local statistics have been used in both a confirmatory manner, to test hypotheses, and in an exploratory manner, where the intent is more to suggest, rather than confirm, hypotheses.

Local statistics may be used to detect clusters either when the location is prespecified (focused tests) or when there is no *a priori* idea of cluster location. When a global test finds no significant deviation from randomness, local tests may be useful in uncovering isolated hotspots of increased incidence. When a global test does indicate a significant degree of clustering, local statistics can be useful in deciding whether (a) the study area is relatively homogeneous in the sense that local statistics are quite similar throughout the area, or (b) there are local outliers that contribute to a significant global statisitc. Anselin (1995) discusses local tests in more detail.

8.4.2 Local Moran Statistic

The local Moran statistic is

$$I_i = n(y_i - \bar{y}) \sum_{j \neq i} w_{ij}(y_j - \bar{y}) \tag{8.27}$$

The sum of local Morans is equal to, up to a constant of proportionality, the global Moran; i.e., $\sum I_i = I$. For example, the local Moran statistic for region 1 in Figure 8.7 is

$$I_1 = (32 - 21)[(26 - 21) + (19 - 21)] = 33 \tag{8.28}$$

The expected value of the local Moran statistic is

$$E[I_i] = \frac{-\sum_{j \neq i} w_{ij}(y_j - \bar{y})}{n - 1} \tag{8.29}$$

and the expression for its variance is more complicated. Anselin gives the variance of I_i, and assesses the adequacy of the assumption that the test statistic has a normal distribution under the null hypothesis.

8.4.3 *Getis's G_i^* Statistic*

To test whether a particular location i and its surrounding regions constitute a cluster of higher (or lower) than average values on a variable (x) of interest, Ord and Getis (1995) have used the statistic

$$G_i^* = \frac{\sum_j w_{ij}(d)x_j - W_i^*\bar{x}}{s\{[nS_{1i}^* - W_i^{*2}]/(n-1)\}^{1/2}} \tag{8.30}$$

where s is the sample standard deviation of the x values, and $w_{ij}(d)$ is equal to one if region j is within a distance of d from region i, and 0 otherwise. The sum is over all regions, including region i. Also,

$$\left.\begin{aligned} W_i^* &= \sum_j w_{ij}(d) \\ S_{1i}^* &= \sum_j w_{ij}^2 \end{aligned}\right\} \tag{8.31}$$

Ord and Getis note that when the underlying variable has a normal distribution, so does the test statistic. Furthermore, the distribution is asymptotically normal even when the underlying distribution of the x-variables is not normal, if the distance d is sufficiently large. Since the statistic (8.30) is written in standardized form, it can be taken as a standard normal random variable, with mean 0 and variance 1.

For region 1 in Figure 8.7, we will use weights equal to 1 for regions 1, 2, and 3, and weights equal to 0 for other regions. The G_i statistic is

$$G_1^* = \frac{77 - 3(21)}{6.69\sqrt{(6(3) - 9)/5}} = 1.56 \tag{8.32}$$

Since this variable has a normal distribution with mean 0 and variance 1 under the null hypothesis that region 1 is not located in a region of particularly high values, we can use a one-sided test with $\alpha = 0.05$ and $z = 1.645$. We therefore fail to reject the null hypothesis.

8.5 Finding Moran's *I* Using *SPSS for Windows 9.0*

Consider the six-region system in Figure 8.7. With connectivity defined by a binary 0–1 weight for adjacent regions, we have the weight matrix given by Equation 8.21. To compute the value of Moran's I in *SPSS*, we first convert the six regional values to z-scores. For the six regions, the z-scores are 1.64, 0.747, −0.299, −0.448, −0.598, and −1.046. Then the quantities $a_i = \sum_j w_{ij} z_j$ are found. These are simply weighted sums of the z-scores of the regions that i is connected to. For example, region 1 is connected to region 2 and 3. For region 1, $a_1 = 0.747 - 0.299 = 0.448$. The six a_i scores are 0.448, 0.299, 0.747, 0.149, −1.046, and −0.896. Now perform a regression, using the as as the dependent variable and the zs as the independent variable. In *SPSS*, click on Analyze, Regression, Linear, and define the dependent and independent variables. Then, under Options, make sure the box labeled "Include constant in equation" is NOT checked. This yields a regression coefficient of 0.446 for the numerator.

For the denominator, we again use no-intercept regression to regress six y-values on six x-values. The six "y-values" are the sums of the weights in each row (2, 4, 4, 2, 4, and 2 for rows 1–6, respectively). The six x-values are 1, 1, 1, 1, 1, and 1 (this will always be a set of n ones, where n is the number of regions). After again making sure that a constant is NOT included in the regression equation, one finds the regression coefficient is 3.0. Moran's I is simply the ratio of these two coefficients: $0.446/3 = 0.1487$.

The variance of I in this example may be found from Equation 8.25:

$$V[I] = \frac{2(36)(5)(18) - 4(6)(5)(60) + 2(4)(18)^2}{7(5)^2(18)^2} = 0.033$$

The z-value associated with a test of the null hypothesis of no spatial autocorrelation is $(0.1487 - (-0.2)/\sqrt{0.033} = 1.92$. This would exceed the critical value of 1.645 under a one-sided test (which we would use, for example, if our initial alternative hypothesis was that positive autocorrelation existed), and would be slightly less than the critical value of 1.96 in a two-sided test. We note, however, that we are on shaky ground in assuming that this test statistic has a normal distribution, since the number of regions is small. We also note that, in this case, the approximation of $1/(3n)$ described in Section 8.3.3 for the variance of I would have yielded a variance of $1/18 = 0.0555$, which is not too far from that found above using Equation 8.25. The approximation of two divided by the sum of the weights, also described in Section 8.3.3, would have yielded $2/18 = 0.1111$. This approximation works better for systems with a greater number of regions.

Exercises

1. The following residuals are observed in a regression of wheat yields on precipitation and temperature over a 6-county area:

County:	1	2	3	4	5	6	
+	7	10	12	9	14	15	Number of positive residuals
–	12	8	19	10	10	10	Number of negative residuals

Use the chi-square test to determine whether there is any interaction between location and the tendency of residuals to be positive or negative. If you reject the null hypothesis of no pattern, describe how you might proceed in the regression analysis.

2. A regression of sales on income and education leaves the following residuals:

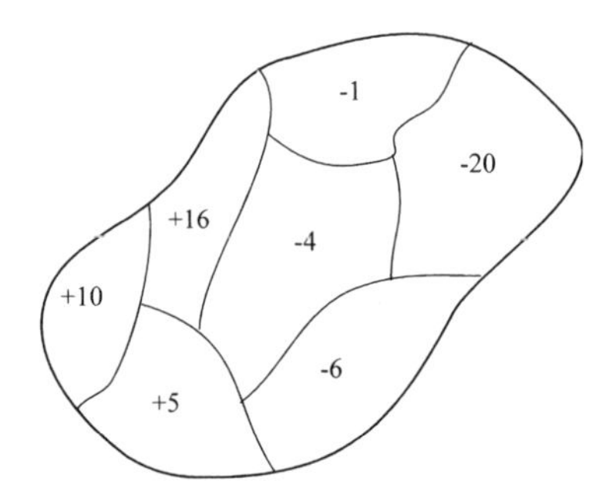

(a) Use the join count statistic to determine whether there is a spatial pattern to the residuals.
(b) Use Moran's I to determine whether there is a spatial pattern to the residuals.
(c) If you reject the null hypothesis in either (a) or (b), describe how you would proceed with the regression analysis.

3.

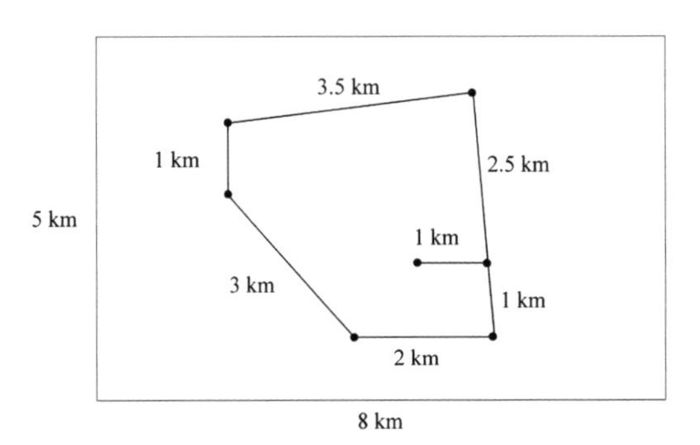

(a) Find the nearest neighbor statistic for the following pattern:

(b) Test the null hypothesis that the pattern is random by finding the z-statistic: $z = 3.826(R_0 - R_e)\sqrt{\lambda n}$, where n is the number of points and λ is the density of points.

(c) Find the chi-square statistic, $\chi^2 = (m-1)s^2/\bar{x}$ for a set of 81 quadrats, where 1/3 of the quadrats have 0 points, 1/3 of the quadrats have 1 point, and 1/3 of the quadrats have 2 points. Then find the z-value to test the hypothesis of randomness, where

$$z = \frac{\chi^2 - (m-1)}{\sqrt{2(m-1)}}$$

where m is the number of cells. Compare it with a critical value of $z = -1.96$ and $z = +1.96$.

4. Find the expected and observed number of black–white joins in the following pattern:

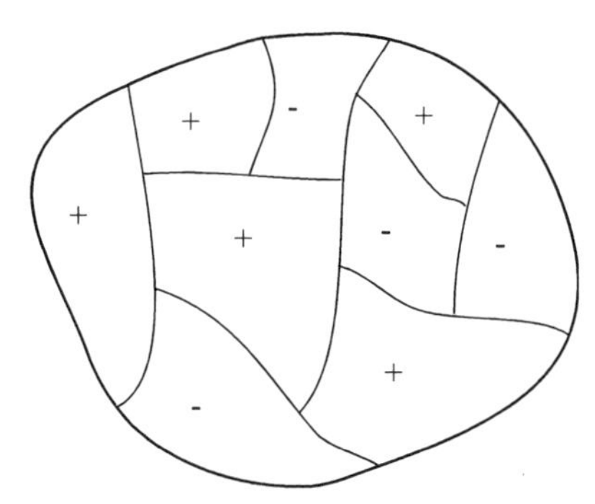

On the basis of your answer, in which direction away from random would you describe this pattern – more toward a checkerboard pattern, or more toward a clustered pattern?

5. Vacant land parcels are found at the following locations:

Find the variance and mean of the number of vacant parcels per cell, and use the variance–mean ratio to test the hypothesis that parcels are distributed randomly (against the two-tailed hypothesis that they are not).

6. Find the nearest neighbor statistic (the ratio of observed to expected mean distances to nearest neighbors) when n points are equidistant from one another on the circumference of a circle with radius r, and there is one additional point located at the center of the circle. (Hints: the area of a circle is πr^2 and the circumference of a circle is $2\pi r$.)

7. Prove that the following two z-scores are equivalent:

$$\frac{R-1}{\sigma_R} = \frac{r_0 - r_e}{\sigma_r}$$

where

$$\sigma_R = 0.52/\sqrt{n}; \qquad \sigma_r = \frac{0.26}{\sqrt{n\rho}}; \qquad R = r_0/r_e$$

and r_0 and r_e are the observed and expected distances to nearest neighbors, respectively. Thus there are two equivalent ways of carrying out the nearest neighbor test.

9 Some Spatial Aspects of Regression Analysis

LEARNING OBJECTIVES

- How to include spatial considerations into regression analyses
- Added-variable plots for spatial variables
- Spatial regression analysis
- Spatially varying parameters, including the expansion method and geographically weighted regression

9.1 Introduction

We have already noted that spatial autocorrelation presents difficulties in estimating regression relationships. In some cases, we may be interested in the pattern of spatially correlated residuals for its own sake. Figure 9.1 is a map I produced for an undergraduate project, showing the residuals from a regression of snowfall on temperature, elevation, and latitude. In this case, the primary purpose was to obtain a visual impression of the effect of the North American Great Lakes on snowfall patterns in New York State. One can clearly see two bands of excess snowfall, one downwind from Lake Erie, and the other downwind from Lake Ontario. The effects downwind of Lake Erie are particularly strong, ranging up to 50–60 inches a year greater than that predicted by temperature, elevation, and latitude alone. The remainder of the map has relatively small residuals. One might also speculate that the negative residuals along the northeast border of the state constitute a precipitation shadow effect, since this area is directly east of the Adirondack Mountains and much of the moisture would have precipitated out before reaching the eastern border.

In the snowfall example, it was not necessary to have precise estimates of the effects of temperature, elevation, and latitude on snowfall, since primary interest was in the map pattern of the residuals. However, spatial autocorrelation in the residuals violates an underlying assumption of ordinary least-squares regression, and so alternatives must be considered when reliable regression equations are desired. Spatial regression models seek to remedy the situation by adding to the list of explanatory variables the values of x and/or y in surrounding regions as well. Some approaches to these spatial regression models are considered in Sections 9.2 and 9.3.

Up to this point, we have assumed that values of the regression coefficients were global, in the sense that they were thought to apply to the region as a

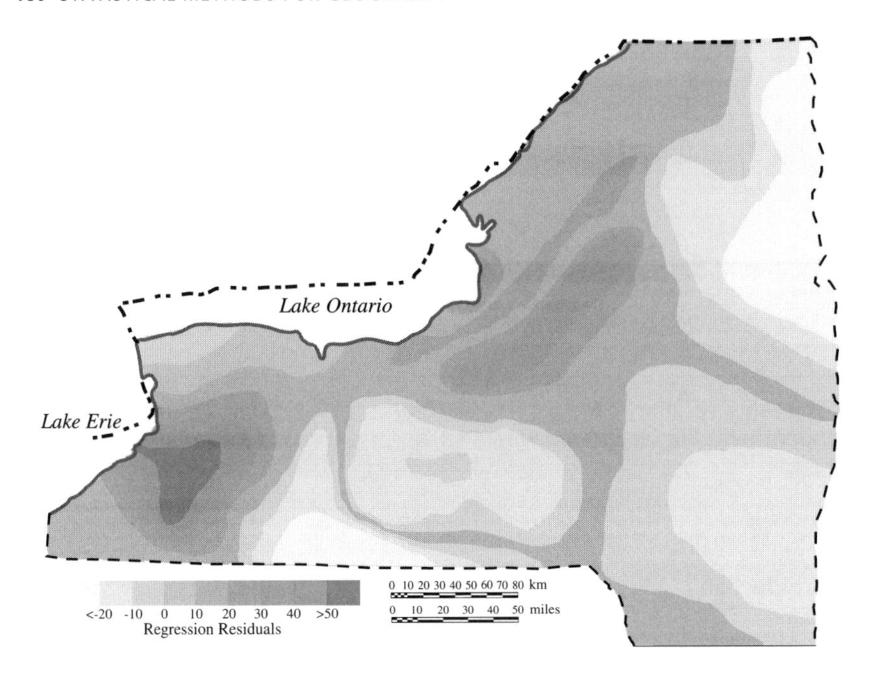

Figure 9.1 **Regression residuals from snowfall analysis**

wholc. Howcvcr, it is possiblc that thc cocfficicnts vary ovcr spacc. Section 9.4 examines two approaches to spatially varying regression parameters. The final section provides an illustration of the various methods.

9.2 Added-Variable Plots

When regression residuals exhibit spatial autocorrelation, this suggests that the regression results may benefit from additional explanatory variables. Haining (1990b) notes that *added variable plots* are "graphical devices that are used to decide whether a new explanatory variable should be added to a regression" (see also Weisberg 1985, Johnson and McCulloch 1987). He identifies four situations where spatial effects may be entered into the right-hand side of a regression equation:

(1) the value of y depends upon values of y nearby;
(2) the value of y at a site depends not only upon values of x at the site but also upon values of x at nearby sites;
(3) the value of y at a site depends upon the value of x at the site and on values of x and y at nearby sites; and
(4) the size of the error at a site is related to the size of the error at nearby sites.

Case (4) is statistically indistinguishable from case (3).

The idea behind added variable plots is to see whether there is a relationship between y, once it has been adjusted for the variables already in the equation, and some omitted variable. Let x_p denote the omitted variable. The procedure is as follows:

(1) Obtain the residuals of the regression of y on the x-variables.
(2) Obtain the residuals of the regression of x_p on the x-variables.
(3) Plot the residuals obtained in (1) on the vertical axis, and those from (2) on the horizontal axis.

The result is the relationship between x_p and y, adjusted for the other xs. If the points in the plot lie along or near a straight line, this suggests that the variable should be added to the regression equation. These plots may be produced within SPSS by checking the "Produce all Partial Plots" box under the Plots section of Linear Regression.

9.3 Spatial Regression

It is possible to specify a spatial regression model in the same way as the usual linear regression model, with the exception that the residuals are modeled as functions of the surrounding residuals (see, e.g., Bailey and Gatrell 1995). If we use ε to denote the usual residual or error term, the residual for a particular observation is written as a linear function of the other residuals:

$$\varepsilon_i = \rho \sum_{j=1}^{n} w_{ij}\varepsilon_j + u_i \tag{9.1}$$

where w_{ij} is a measure of the connection between location i and location j (often taken as a binary connectivity measure), ρ is a measure of the strength of the correlation of the residuals, and u_i is the remaining error term after the correlation among residuals has been accounted for. Note that if $\rho = 0$, the model reduces to the ordinary linear regression model.

To estimate the model, one can define the quantities

$$\left.\begin{aligned} y_i^* &= y_i - \rho \sum_{j=1}^{n} w_{ij} y_j \\ x_i^* &= x_i - \rho \sum_{j=1}^{n} w_{ij} x_j \end{aligned}\right\} \tag{9.2}$$

Then regressions of y^* vs x^* are tried for a variety of ρ values, beginning at zero. The residuals of each regression are inspected, and the value of ρ associated with the most suitable set of residuals is adopted. Section 9.5.3 provides an example. Bailey and Gatrell note that this estimation procedure is, strictly, not one that is the best from a statistical viewpoint, and that more sophisticated approaches exist. However, it should give the analyst a good idea of the spatial effects that may be present in a model.

9.4 Spatially Varying Parameters

9.4.1 The Expansion Method

With linear regression, the slope and intercept parameters are "global", in the sense that they apply to all observations. The expansion method (Casetti 1972, Jones and Casetti 1992) suggests that these parameters may themselves be functions of other variables. Thus, in a linear regression equation of house prices (y) on lot size (x_1) and number of bedrooms (x_2):

$$y = b_0 + b_1 x_1 + b_2 x_2 + \varepsilon \tag{9.3}$$

the effect of lot size on house prices (b_1) may itself depend upon whether there is a park nearby (for example, large lot sizes may be more valuable in a suburb if there is no other green space nearby). So, we add an expansion equation

$$b_1 = c_0 + c_1 d \tag{9.4}$$

where d is the distance to the nearest park. We would expect c_1 to be positive; large distances to the nearest park would mean that b_1 is high, which in turn means that lot sizes have a large influence on house prices.

If we substitute this expansion equation into the original equation we have

$$y = b_0 + (c_0 + c_1 d)x_1 + b_2 x_2 + \varepsilon = b_0 + c_0 x_1 + c_1 d x_1 + b_2 x_2 + \varepsilon \tag{9.5}$$

To estimate the coefficients, we perform a linear regression of y on the variables x_1, x_2, and dx_1. In Equation 9.5, the new quantity dx_1 may be thought of as a new variable, created by multiplying together distance to park (d) and lot size (x_1). When the coefficient c_1 is significant, this is known as an *interaction* effect; the effect of lot size on housing prices interacts with, or depends upon, the distance to the park (or alternatively, the effect of distance to the park depends upon the size of the lot).

The edited collection of Jones and Casetti (1992) contains a wide variety of applications of the expansion method. These include applications to models of welfare, population growth and development, migrant destination choice, urban development, metropolitan decentralization, and the spatial structure of agriculture. The collection also includes methodological contributions that focus upon statistical aspects of the model, including its relationship to spatial dependence in the data.

9.4.2 Geographically Weighted Regression

In a series of articles, Fotheringham and his colleagues at Newcastle have outlined an alternative approach to the expansion method that accounts for spatially varying parameters (see, e.g., Fotheringham *et al.* 1998, Brunsdon *et al.* 1996, 1999). Their geographically weighted regression (GWR) technique is based upon "local" views of regression as observed from any location. For each location, one can estimate a regression equation where weights are attached to observations surrounding the location. Relatively large weights are given to points near the location, and smaller weights are assigned to observations far from the location. As Fotheringham *et al.* (2000) note:

> There is a continuous surface of parameter values... In the calibration of the GWR model it is assumed that observed data near to point *i* have more of an influence in the estimation of the [regression coefficients] than do data located farther from *i* (p. 108).

More formally, the dependent variable at location i is modeled as follows:

$$y_i = b_{i0} + \sum_{j=1}^{p} b_{ij}x_{ij} + \varepsilon_i \tag{9.6}$$

where, as is the case with simple linear regression, there are p independent variables, and x_{ij} represents the observation on variable j at location i. The important point to note is that the b coefficients have i subscripts, indicating that they are specific to the location of observation i.

One reasonable choice for the weights is a negative exponential function of squared distance

$$w_{ij} = \mathrm{e}^{-\beta d_{ij}^2} = \exp(-\beta d_{ij}^2) \tag{9.7}$$

so that points that are farther away will be assigned lower weights.

To estimate the regression coefficients at location i, one first defines the weights (w_{ij}), using an initial "guess" for the value of β (one possibility would be to use $\beta = 0$, which corresponds to the ordinary least-squares case). Then define the quantities

$$\left.\begin{aligned} y_j^* &= \sqrt{w_{ij}}y_j \\ x_j^* &= \sqrt{w_{ij}}x_j \quad j = 1, \ldots, n \end{aligned}\right\} \tag{9.8}$$

These are the weighted observations. At location i, run a linear regression of the y^* on the x^*, omitting observation i from the analysis. Use the resulting regression coefficients to predict the value of y at location i. Then find the squared difference between the observed value of y (denoted y_i) and this predicted value

$$\{y_i - \hat{y}_{\neq i}(\beta)\}^2 \tag{9.9}$$

where $\hat{y}_{\neq i}(\beta)$ is the predicted value of the dependent variable at location i when observation i has not been used in the estimation, and the β reminds us that this prediction was made using a specific value of β. After this has been repeated for each location i, one may compute the total sum of squared deviations between observed and predicted values as

$$s(\beta) = \sum_{i=1}^{n} \{y_i - \hat{y}_{\neq i}(\beta)\}^2 \tag{9.10}$$

The next step is to repeat this procedure for many values of β, choosing as "best" the value of β that minimizes the score $s(\beta)$. This final value of β yields the best set of weights. The final regression coefficients at each location are given as follows: first use the final, optimal value of β to define the weights, and then regress y^* and x^* using *all* of the observations.

9.5 Illustration

Figure 9.2 displays the location of 30 hypothetical houses in a square study area that features a park at its center. The dataset in Table 9.1 was generated by assuming that housing prices were related to lot size, number of bedrooms, and the presence of a fireplace. Furthermore, spatial effects were added in the generation of the data. The lot sizes were generated in such a way as to be spatially autocorrelated, and the effect of lot size on housing prices was made to be a function of how distant the house was from the centrally located park.

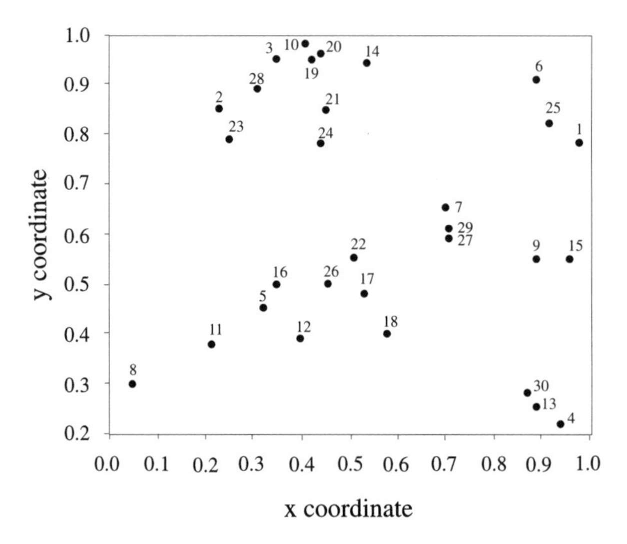

Figure 9.2 **Location of thirty hypothetical houses**

Table 9.1 **Hypothetical data on 30 houses**

CASE	XCOORD	YCOORD	PRICE	LOTSIZE	BEDRMS	FIREPLC
1	.9619	.7817	224323	5.987	3	1
2	.2378	.8520	143510	4.241	2	1
3	.3481	.9440	233533	5.039	6	1
4	.9329	.2235	192328	3.100	4	1
5	.3258	.4532	158553	5.133	2	0
6	.8847	.9136	297893	6.397	5	1
7	.7063	.6176	150054	7.590	2	1
8	.0473	.2902	193785	4.848	6	0
9	.8927	.5538	206744	4.272	6	0
10	.4131	.9766	159585	5.126	3	0
11	.2189	.3649	212046	4.583	5	0
12	.3957	.3827	171795	4.589	3	1
13	.8909	.2550	125737	2.343	3	0
14	.5363	.9402	253078	7.138	4	1
15	.9574	.5488	189896	5.016	4	0
16	.3571	.5017	228830	6.783	5	1
17	.5396	.4733	163033	8.169	2	1
18	.5687	.3996	202935	5.199	2	1
19	.4256	.9444	205478	6.426	3	0
20	.4431	.9568	207324	6.199	2	1
21	.4555	.8451	249965	7.895	3	0
22	.5191	.5430	193800	8.689	4	0
23	.2518	.7851	203844	5.000	5	0
24	.4458	.7717	153122	5.654	2	0
25	.9242	.8261	252367	6.959	4	0
26	.4457	.4998	101089	5.376	2	0
27	.7138	.5867	196954	6.806	4	0
28	.3222	.8879	158972	4.352	2	1
29	.7107	.6137	191339	8.170	4	0
30	.8657	.2810	184990	3.377	4	0

More specifically, housing prices were generated using the equations

$$\left.\begin{aligned} p &= 20\,000 + b_1x_1 + 20\,000x_2 + 20\,000x_3 + \varepsilon \\ b_1 &= 10\,000 + 20\,000d \\ \varepsilon &\sim N(0,\, 20\,000^2) \end{aligned}\right\} \tag{9.11}$$

All digits have been retained in the generated prices, though in practice one would expect them to be rounded to, say, the nearest hundred. Where p is price, x_1 is lot size (in thousands of square feet), x_2 is number of bedrooms, and x_3 is a dummy variable indicating the presence or absence of a fireplace ($1 = $presence; $0 = $absence). d is the distance from the centrally located park, and ε is a normally distributed error term with standard deviation equal to 20 000. Houses were assigned fireplaces with probability 0.3, and were assigned a number of bedrooms by allowing integers in the range 2–6 with equal likelihoods. Lot sizes were normally distributed with mean 6 and standard deviation 0.8.

Thus the "true" data follow quite closely an expansion-equation model, and we will expect that such a model will perform quite well. But for now, let us assume that we are simply faced with the data in Table 9.1, and we want to model housing prices as a function of the independent variables.

9.5.1 Ordinary Least-Squares Regression

Table 9.2 shows the results from the ordinary least-squares regression of housing price on lot size, number of bedrooms, and presence of fireplace. The coefficients are all significant. The r^2 value is 0.562, and the standard error of the estimate is 29 080. The residuals display positive spatial autocorrelation, indicating potential problems with the estimation. The coefficient on lot size is a bit low, since we know from the way the data were generated that it ranges from a low of 10 000 near the park to a high of about 20 000 ($=10\,000+20\,000(0.5)$) near the periphery.

9.5.2 Added-Variable Plots

We begin by making the rather arbitrary decision that neighbors are defined in this example as the three closest observations. Thus $w_{ij}=1$ if observation j is one of the three nearest neighbors of i, and 0 otherwise.

Table 9.2 **Results from OLS regression**

Variables Entered/Removed[b]

Model	Variables Entered	Variables Removed	Method
1	FIREPLC, LOTSIZE, BEDRMS[a]		Enter

[a]All requested variables entered.
[b]Dependent Variable: PRICE

Model Summary

Model	R	R Square	Adjusted R Square	Std. Error of the Estimate
1	.749[a]	.562	.511	29080.13

[a]Predictors: (Constant), FIREPLC, LOTSIZE, BEDRMS

ANOVA[b]

Model		Sum of Squares	df	Mean Square	F	Sig.
1	Regression	2.8E+10	3	9.4E+09	11.108	.000[a]
	Residual	2.2E+10	26	8.5E+08		
	Total	5.0E+10	29			

[a]Predictors: (Constant), FIREPLC, LOTSIZE, BEDRMS
[b]Dependent Variable: PRICE

Coefficients[a]

Model		Unstandardized Coefficients		Standardized Coefficients	t	Sig.
		B	Std. Error	Beta		
1	(Constant)	51249.856	26866.791		1.908	.068
	LOTSIZE	10323.904	3444.409	.391	2.997	.006
	BEDRMS	20628.873	4153.976	.661	4.966	.000
	FIREPLC	24834.295	10942.393	.301	2.270	.032

[a]Dependent Variable: PRICE

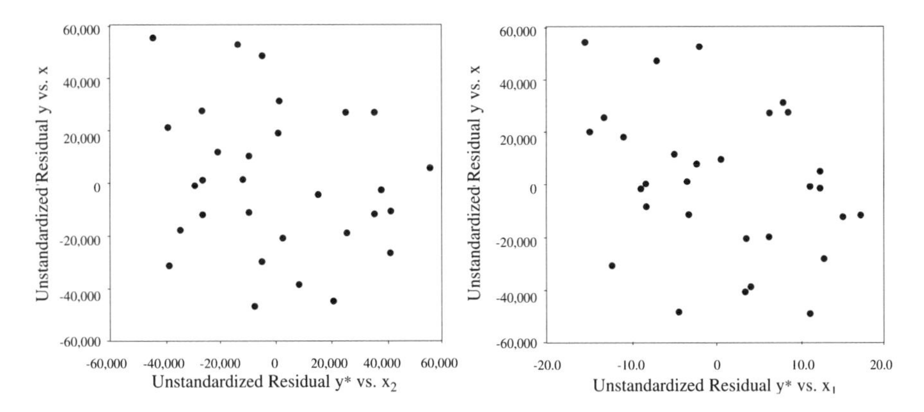

Figure 9.3 **Added variable plots**

Following Haining's example, we will consider the addition of new variables. The possibilities we will consider are

$$\left.\begin{aligned} x^*_{i(1)} &= \sum_{j=1}^{n} w_{ij}y_j \\ y^*_{i(2)} &= \sum_{j=1}^{n} w_{ij}x_j \end{aligned}\right\} \qquad (9.12)$$

The first suggests that y at a location is a function not only of x at that location but also of the y values in surrounding locations. The second equation suggests that the y value at a location may also be a function of the x-values in surrounding locations. To construct added variable plots for each of these potential additions to the regression equation, we need (a) the residuals of the ordinary least-squares regression (from Section 9.5.1), and (b) the residuals from regressions of x^*_i on x. These residual plots are shown in panels (a) and (b) of Figure 9.3. Neither plot shows a significant correlation, and so we conclude that these variables would not improve the specification of the regression equation.

9.5.3 Spatial Regression

Following the autocorrelated errors model of Section 9.3, and using the same definition of the weights (w) used in Section 9.5.2, we define the quantities

$$\left.\begin{aligned} y^*_i &= y_i - \rho\sum_{j=1}^{n} w_{ij}y_j \\ x^*_i &= x_i - \rho\sum_{j=1}^{n} w_{ij}x_j \end{aligned}\right\} \qquad (9.13)$$

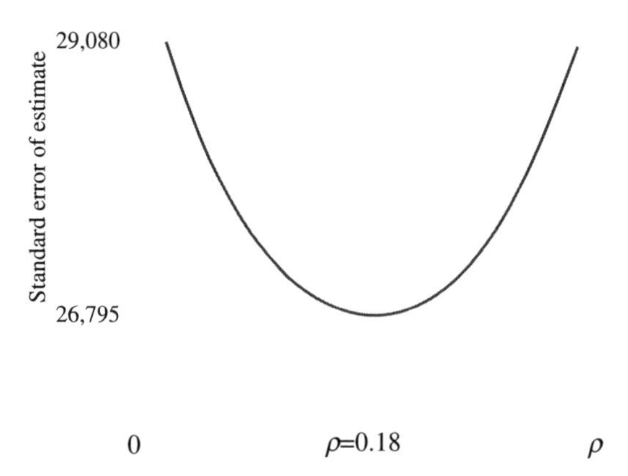

Figure 9.4 **Minimizing the standard error of the estimate**

Table 9.3 **Spatial regression results with $\rho = 0.18$**

	Coefficient	Standard error	t
Intercept	2 926.9	4 914	0.60
Lot size	17 910	3 341	5 30
No. of bedrooms	22 921	3 139	7.30
Fireplace	27 233	9 003	3.02

We would like to choose a value of ρ that is associated with a "good" set of residuals. Although there are different ways that this could be done, after trying different ρ values we find that $\rho = 0.18$ minimizes the standard error of the estimate (see Figure 9.4). When y^* is regressed on x^* using this value of ρ, we obtain the results in Table 9.3. The standard error of the estimate has been reduced to 26 795, and the value of r^2 is now 0.883. All variables are significant as before, and the t-values for all coefficients are higher than those under ordinary least squares (Section 9.5.1). In addition, the coefficient on lot size in this equation is equal to 17 910, which is closer to its average value of about 15 000 (recall that we generated the data so that the true lot size coefficient varied from 10 000 to about 20 000).

9.5.4 Expansion Method

Next we estimate the expansion model

$$\left.\begin{aligned} p &= b_0 + b_1x_1 + b_2x_2 + b_3x_3 + \varepsilon \\ b_1 &= \gamma_0 + \gamma_1 d \end{aligned}\right\} \qquad (9.14)$$

where the variables are as defined above, and γ_0 and γ_1 are the regression coefficients that tell us how the influence of lot size on housing prices varies with distance from the park. This may be rewritten as

$$p = b_0 + (\gamma_0 + \gamma_1 d)x_1 + b_2 x_2 + b_3 x_3 + \varepsilon \tag{9.15}$$

which is identical to

$$p = b_0 + \gamma_0 x_1 + \gamma_1 d x_1 + b_2 x_2 + b_3 x_3 + \varepsilon \tag{9.16}$$

The results obtained when using ordinary least-squares regression on this equation are shown in Table 9.4.

The r^2 value is equal to 0.747, and all parameters, including those associated with the expansion equation, are significant. Furthermore, all parameter values are near their "true" values, and the standard error of the estimate is 22 552.

Of course, it should be kept in mind that one reason this particular approach has worked relatively well here is that the estimated model is consistent with the way in which the data were generated. We helped ourselves out by choosing to expand the model using the relation between lot size effects and distance

Table 9.4 **Results from expansion method**

Variables Entered/Removed[b]

Model	Variables Entered	Variables Removed	Method
1	DPLOT FIREPLC, LOTSIZE, BEDRMS[a]		Enter

[a]All requested variables entered.
[b]Dependent variable: PRICE

Model Summary[b]

Model	R	R Square	Adjusted R Square	Std. Error of the Estimate
1	.864[a]	.747	.706	22552.50

[a]Predictors: (Constant), DPLOT, FIREPLC, LOTSIZE, BEDRMS
[b]Dependent Variable: PRICE

ANOVA[b]

Model		Sum of Squares	Df	Mean Square	F	Sig.
1	Regression	3.7E+10	4	9.4E+09	18.409	.000[a]
	Residual	1.3E+10	25	5.1E+08		
	Total	5.0E+10	29			

[a]Predictors: (Constant), DPLOT, FIREPLC, LOTSIZE, BEDRMS
[b]Dependent Variable: PRICE

Coefficients[a]

Model		Unstandardized Coefficients		Standardized Coefficients		
		B	Std. Error	Beta	t	Sig.
1	(Constant)	42361.839	20939.718		2.023	.054
	FIREPLC	20632.446	8543.021	.250	2.415	.023
	BEDRMS	16482.800	3364.706	.528	4.899	.000
	LOTSIZE	8313.839	2712.410	.315	3.065	.005
	DPLOT	19743.607	4624.268	.454	4.270	.000

[a]Dependent Variable: PRICE

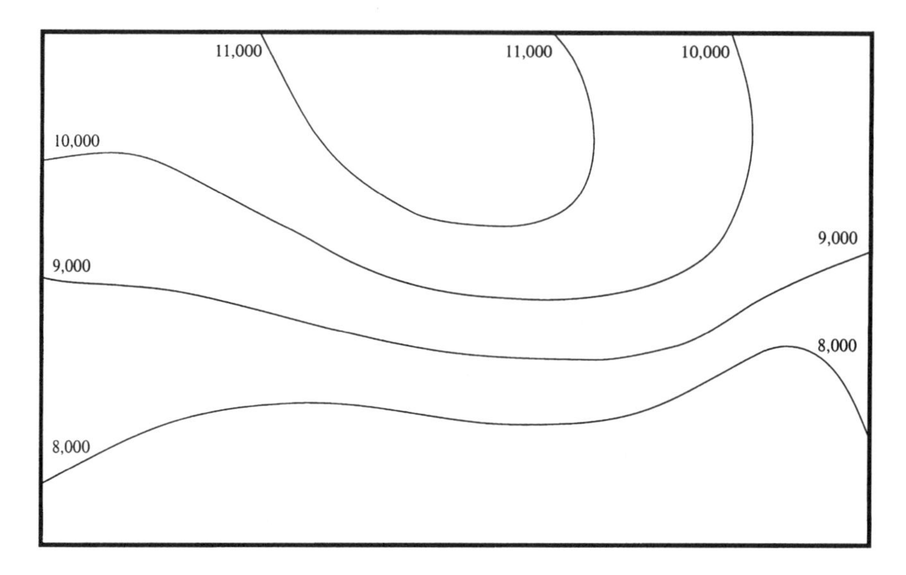

Figure 9.5 **Spatial variation in lot size coefficient**

from the park – this was a good choice because that is how the data were created!

9.5.5 Geographically Weighted Regression

Using the weights defined in Equation 9.7 and the method outlined in Section 9.4.2, the optimal value of β was found to be 4.8. This defines a set of weights that are associated with the variables in Equation 9.8. The regressions are then run once for each data point, using these weights. Figure 9.5 displays a map of the coefficient on lot size. From the figure one can see that the parameter is higher away from the park, where there is a large cluster of observations in the north. This is in keeping with our expectations, since the effect of lot size on house prices was made to be greater in peripheral locations.

Exercises

1. Use an added variable plot to determine whether distance to the park should be added to a regression of housing price on lot size, number of bedrooms, and presence or absence of a fireplace. Use the data in Table 9.1.

2. Using the data in Table 9.1 and geographically weighted regression, produce a map showing the spatial variation in the coefficient on number of bedrooms. Alternatively, you may provide a table showing the regression coefficient for the number of bedrooms at each of the 30 sample locations.

3. With the data in Table 9.1, first perform an ordinary least-squares regression with housing price as the dependent variable and lot size as the independent variable. Then use the expansion method, with the lot size coefficient depending upon the number of bedrooms. Interpret the results.

4. Use the spatial regression method outlined in Sections 9.3 and 9.5.3 with the data in Table 9.1 for a regression of housing prices on lot size and number of bedrooms.

10 Data Reduction: Factor Analysis and Cluster Analysis

LEARNING OBJECTIVES

- Introduction to multivariate methods for data reduction, including principal components analysis, factor analysis, and cluster analysis
- Geometric interpretations of the methods

10.1 Factor Analysis and Principal Components Analysis

Many studies of complex geographic phenomena begin with a set of data and notions of hypotheses and theories that are vague at best. Factor analysis may be used as a data reduction method, to reduce a dataset containing a large number of variables down to one of more manageable size. When many of the original variables are highly correlated, it is possible to reduce the original data from a large number of original variables to a small number of underlying factors.

A geometric interpretation helps one to understand the purpose of factor analysis. A data set consisting of n observations on p variables may be represented as n points plotted in a p-dimensional space. This is easiest to imagine when $p = 1$, 2, or 3, and the latter case is illustrated in Figure 10.1. The figure also shows an ellipsoidal figure that contains the majority of the data points. The idea behind factor analysis is to construct factors that represent a large proportion of the variability of the dataset. The first factor is, geometrically, the longest axis of the ellipse. The original axes correspond to variables; the longest axis of the ellipse is a *new* variable, which is a linear combination of the original variables. This new variable, or factor, captures as much of the variability in the dataset as possible.

A second factor is derived by finding the second longest axis of the ellipse, such that this second axis is perpendicular to the first axis. The fact that the axes of the ellipse are perpendicular implies that the newly defined factors will be uncorrelated with one another – they represent separate and independent aspects of the underlying data.

A dataset characterized by an extremely elongated ellipse would be well represented by a single factor – that combination of variables would explain almost all of the variability in the original data. In the extreme case, the plotted data would fall along a single line, which would constitute the axis or single factor that would capture *all* of the variability in the data. At the other

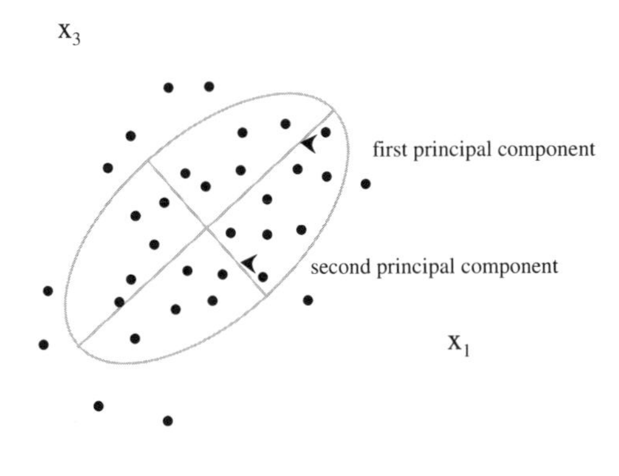

Figure 10.1 **Data ellipsoid in $p = 3$ dimensions**

extreme, the data ellipse could be circular; in this case, all factors explain an equal amount of the variability in the original data, and there are no dominant factors.

In this discussion we will focus more upon the interpretation of the outputs of factor analysis, and less on its mathematical aspects. The next subsection addresses the interpretation of factor analysis results through an example using 1990 census data from Erie County, New York.

10.1.1 Illustration: 1990 Census Data for Buffalo, New York

Geographers often use many census variables in their analyses, and the set of variables can easily contain subsets that measure essentially the same phenomenon. The following example illustrates, for a small set of census data, how the number of original variables can be collapsed into a smaller number of uncorrelated factors.

A 235×5 data table was constructed by collecting and deriving the following information for the 235 census tracts in Erie County, New York (variable labels are in parentheses):

(a) median household income (medhsinc)
(b) percentage of households headed by females (female)
(c) percentage of high school graduates who have a professional degree (educ)
(d) percentage of housing occupied by owner (tenure)
(e) percentage of residents who moved into their present dwelling before 1959 (lres)

These five variables capture different aspects of the socioeconomic and demographic character of census tracts. Do they represent separate dimensions of socioeconomic and demographic structure, or is there significant redundancy in what they measure, indicating that the variables might be reduced to a smaller number of underlying indices or factors?

A natural place to start is with the correlation matrix. Table 10.1 reveals that the highest correlations are with the median household income variable; areas of high income have low percentages of households headed by females, high percentages of homeowners, high percentages of graduates with professional degrees, and a relatively low proportion of long-term residents. Using the test of significance described in Chapter 5, all correlations with absolute value greater than $2/\sqrt{235} = 0.130$ are significant.

The second step is to examine the outcome of describing the data as an ellipsoid, as described above. The method of *principal components* is used to describe the p axes of the ellipse (which in turn is constructed in a p-dimensional space, where p is the number of variables). The relative lengths of the axes are called *eigenvalues*. They are referred to in Table 10.2 as "extraction sums of squared loadings." A "loading" is the correlation between a component or factor and the original variable. If one were to sum the squared correlations between a factor and all of the original variables, this would be equal to the eigenvalue or the length of the ellipse axis. From the table, we see that the highest eigenvalue is 2.6 and the second highest is 0.96. Note that the column displaying these values sums to five – the eigenvalues (i.e., "extraction sums of squared loadings") will always sum to the number of variables. In the extreme case there would be a single component with perfect correlations with all of the original variables. The eigenvalue for this component would be equal to $1^2 + 1^2 + \cdots + = p$. All of the other eigenvalues would be equal to zero, and the ellipse would collapse to a single line.

Table 10.1 **Correlation among variables**

Correlation Matrix

		MEDHSINC	FEMALEH	EDUC	TENURE	LRES
Correlation	MEDHSINC	1.000	-.595	.415	.569	-.455
	FEMALEH	-.595	1.000	-.348	-.531	.221
	EDUC	.415	-.348	1.000	.117	-.161
	TENURE	.569	-.531	.117	1.000	-.438
	LRES	-.455	.221	-.161	-.438	1.000

Table 10.2 **Variance explained by each component**

Total Variance Explained

	Extraction Sums of Squared Loadings			Rotation Sums of Squared Loadings		
Component	Total	% of Variance	Cumulative %	Total	% of Variance	Cumulative %
1	2.602	52.032	52.032	1.035	20.707	20.707
2	.957	19.149	71.181	1.032	20.637	41.344
3	.741	14.826	86.007	1.018	20.358	61.702
4	.362	7.244	93.251	1.005	20.110	81.812
5	.337	6.749	100.000	.909	18.188	100.000

Extraction Method : Principal Component Analysis.

This table also provides us with valuable information concerning how many factors are necessary to adequately describe the data. There are two "rules of thumb" that are used to decide on the number of factors. One such rule is to retain components with eigenvalues greater than one. This would be an unfortunate rule to apply in this instance, since the second eigenvalue is just slightly less than one (0.96). An alternative is to plot the eigenvalues on the vertical axis and the factor number (ranging from 1 to p) on the horizontal axis of a graph. Then inspect the graph to locate a point at which the graph (termed a *scree plot*) flattens out; such a feature implies that the additional factors do not contribute much to the explanation of variability in the data set. Figure 10.2 displays a scree plot for our present example. Some judgement is called for, and we could in this instance justify the extraction of either two or three factors.

Suppose that we decide to extract two factors. The next step is to inspect the loadings, or correlations between the factors and the original variables. This is a key step in the analysis, since it is where the "meaning" and interpretation of each factor occurs. To aid in this interpretation, the extracted component solution is rotated in the p-dimensional space, so that the loadings tend to be either high in absolute value (near plus or minus one) or low (near zero).

Table 10.3 shows that the two-factor solution may be described as follows. The first factor is one where income, tenure, and length of residence all "load highly". We can think of these variables has being combined to form a single index (the factor) that describes with a single number what the three variables represent. The second factor is associated with the other two variables – education and family structure. It is common practice to attempt to give the factors snazzy, descriptive names. Having noted this, it is often difficult to come up

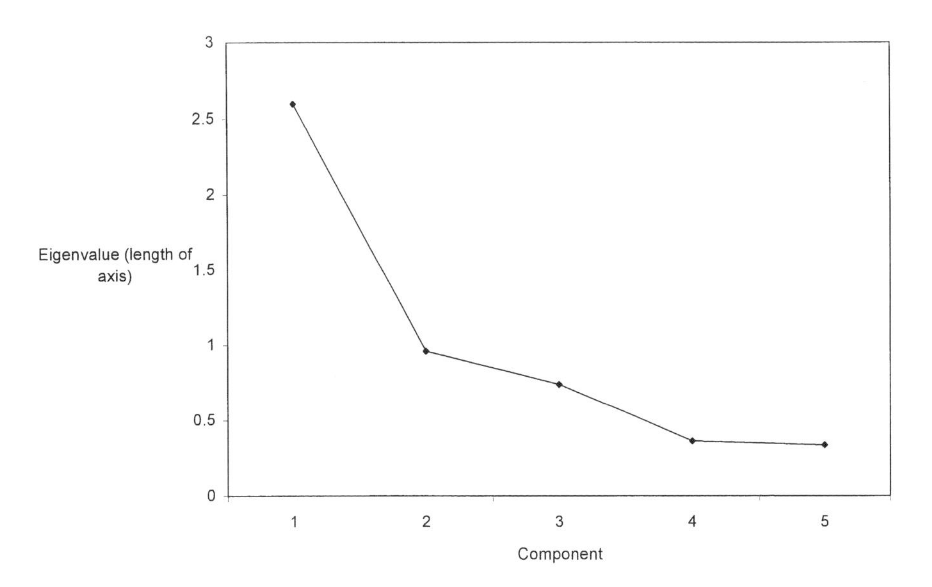

Figure 10.2 **Scree plot for Erie County example**

Table 10.3 **Factor loadings**

Rotated Component Matrix[a]

	Component	
	1	2
MEDHSINC	.668	.562
FEMALEH	−.515	−.608
EDUC	−3.79E-02	.912
TENURE	.848	.154
LRES	−.766	−2.78E-03

Extraction Method : Principal Component Analysis.
Rotation Method: Varimax with Kaiser Normalization.
[a] Rotation converged in 3 iterations.

Table 10.4 **Communalities**

Communalities

	Extraction
MEDHSINC	.762
FEMALEH	.635
EDUC	.833
TENURE	.742
LRES	.587

Extraction Method : Principal Component Analysis.

with something creative! The first factor here might be thought of as a housing/economic factor and the second a sociological factor.

The difference between principal components analysis and factor analysis may be summarized as follows. Principal components analysis is a descriptive method of decomposing the variation among a set of p original variables into p components. The components are linear combinations of the original variables. It is used as a prelude to factor analysis, which attempts to model the variability in the original set of variables via a reduced number of factors which is less than p. In factor analysis, values of the original variables may be reconstructed by writing them as linear combinations of the factors, plus a "uniqueness" term. Alternatively stated, in factor analysis part of the variability in an original variable is captured by the factors (this portion of the variability is termed the communality), and part is *not* captured by the factors (this portion is termed the uniqueness). Table 10.4 shows the communalities for the two-factor solution. The highest communality is for education (0.833) and the lowest for length of residence (0.587). The communalities for a variable are equal to the sum of the squared correlations of the variable with the factors. For example, the communality for education is equal to its squared correlation with factor one (0.0379^2) plus its squared correlation with factor two (0.912^2). Length of residence has the highest uniqueness, since it is not highly correlated with the two factors.

It is important to realize that the output of a factor analysis is a strong function of the input. The fact that length of residence is not strongly related to either factor does not mean that it is not an important feature of urban structure. The factors that emerge from a factor analysis are not necessarily the "most important" ones, but rather the ones that capture the nature of the dataset. If we had a dataset with fifteen variables, and eleven of the fifteen variables were alternative measures of income, we could be certain that an income factor would emerge as the principal factor, simply because so many variables were highly intercorrelated.

Finally, one of the outputs from factor analysis is the *factor scores*. Instead of making p separate maps describing the spatial pattern of each variable, one is now interested in making a number of maps equal to the number of underlying factors. For each factor, and for each observation, a score may be computed as a linear combination of the original variables. The result is a new data table; instead of the original $n \times p$ table, we now have a $n \times k$ table, where k is the number of factors. Figures 10.3 and 10.4 display the factor scores on each of our two factors for the Erie County census tracts.

10.1.2 Regression Analysis on Component Scores

As we have seen, the chief use of principal components analysis is to summarize a large number of variables in terms of a set of uncorrelated components. This sounds ideal for regression analysis, where one is faced with a large number of possibly correlated variables, and the objective is to use a small number of uncorrelated variables that will be useful in explaining the variability in the dependent variable.

Regression analysis may be carried out on component scores, ensuring not only that the independent variables are a parsimonious subset capturing the underlying dimensions of the full set of potential independent variables, but that they are uncorrelated as well. This idea for eliminating multicollinearity is one that is quite commonly employed (for example, see Ormrod and Cole 1996, Ackerman 1998, O'Reilly and Webster 1998). One disadvantage is that it is somewhat more difficult to interpret the regression coefficients. They now indicate how much the dependent variable changes when the component score changes by one unit, and it is more difficult to conceptualize just what a one-unit increase in the component score really implies. Hadi and Ling (1998) also note some pitfalls in the use of principal components regression.

10.2 Cluster Analysis

Whereas factor analysis works by searching for similar *variables*, cluster analysis has as its objective the grouping together of similar *observations*. Since it is conventional to represent each observation as a row in a data table, and each variable as a column, cluster analysis has at its core the search for similar rows of data. Factor analysis is based upon similarities among columns of data.

Like factor analysis, cluster analysis may be thought of as a data reduction technique. We seek to reduce the n original observations into g groups, where

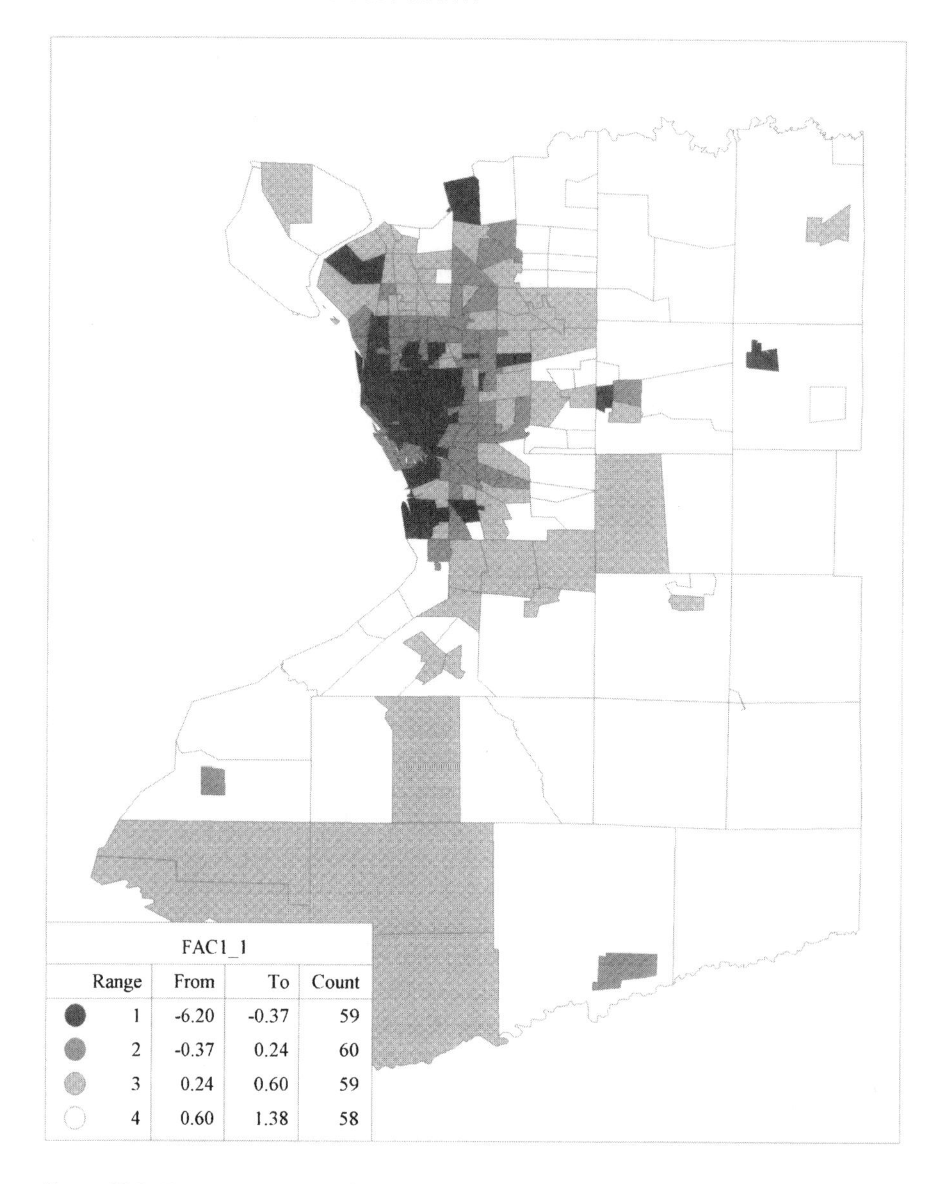

Figure 10.3 **Factor 1 scores**

$1 \leq g \leq n$. In achieving this reduction of n observations into a smaller number of groups, a general goal is to minimize the within-group variation and maximize the between-group variation. In Figure 10.5, there is relatively little variability within groups, as measured by the variation in the location of points around their group centroids. Relative to this within-group variability, there is much more variation in the locations of the group centroids in relation to the centroid for the entire dataset.

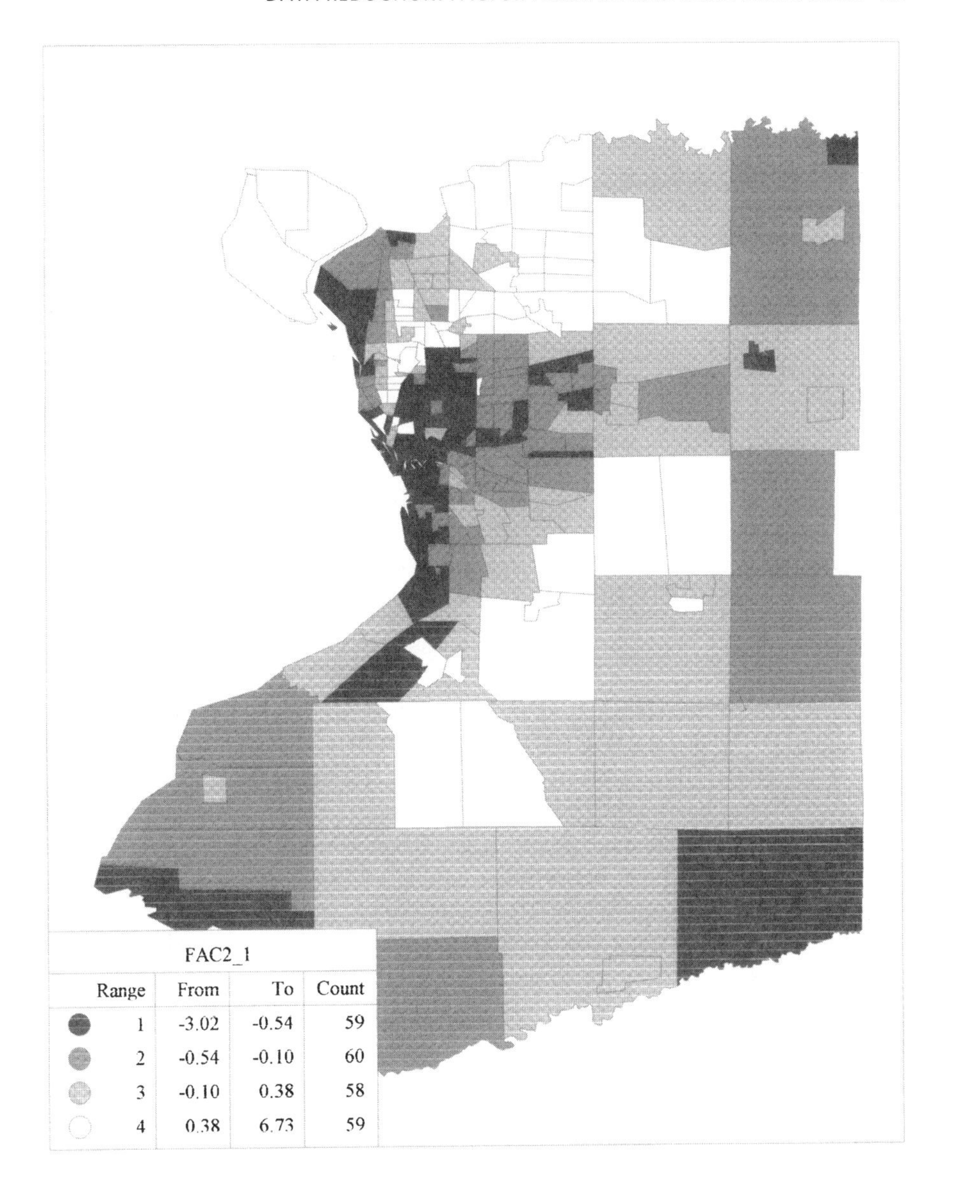

Figure 10.4 **Factor 2 scores**

One of the more widespread applications of cluster analysis in geography has been in the area of geodemographics, where analysts seek to reduce a large number of subregions (e.g., census tracts) by classifying them into a small number of types (see, e.g., Chapter 10 of Plane and Rogerson 1994). Cluster analysis has also been used as a method of regionalization, where the objective is to divide a region into a smaller number of contiguous subregions. In this

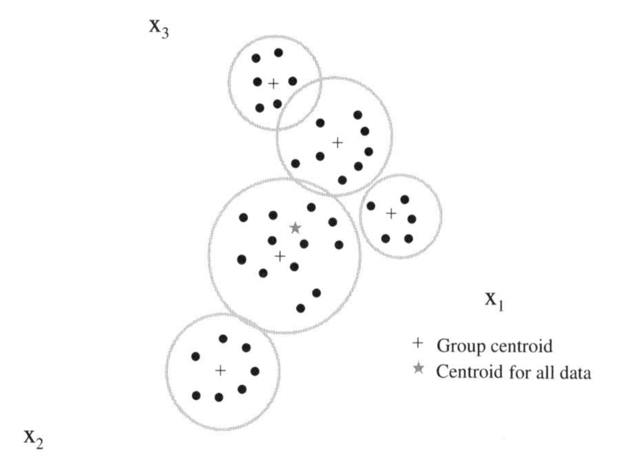

Figure 10.5 **Clustering in $p = 3$ dimensions**

case, it is necessary to modify traditional approaches to cluster analysis slightly to ensure that the created groups are composed of contiguous subregions (see, e.g., Murtagh 1985).

Approaches to cluster analysis may be categorized into two broad types. *Agglomerative* or *hierarchical* methods start with n clusters (where n is the number of observations); each observation is therefore in its own cluster. Then two clusters are merged, so that $n - 1$ clusters remain. This process continues until only one cluster remains (this cluster contains all n observations). The process is hierarchical because the merger of two clusters at any stage of the analysis cannot be undone at later stages. Once two observations have been placed together in the same cluster, they stay together for the remainder of the grouping process.

In contrast, *nonhierarchical* or *nonagglomerative* methods begin with an *a priori* decision to form g groups. Then one begins with either an initial set of g seed points or an initial partition of the data into g groups. If one starts with a set of seed points, a partition of the data into g groups is achieved by assigning each observation to the nearest seed point. If one begins with a partition of the data into g groups, g seed point locations are calculated as the centroids of these g partitioned groups. In either case, an iterative process then takes place, where new seed points are calculated from partitions and then new partitions are created from the seed points. This process continues until no reassignments of observations from one group to another occur. The convergence of this iterative process is usually very rapid.

The nonhierarchical methods have the advantage of requiring less computational resources, and for this reason they are the preferred method when the number of observations is very large. They have the disadvantage that the number of groups must be specified prior to the analysis, though in practice it is not uncommon to find solutions for a range of g values.

10.2.1 More on Agglomerative Methods

With agglomerative methods, at each stage one merges the closest pair of clusters. There are many possible definitions that may be used for "closest". Consider all pairs of distances between elements of cluster A and cluster B. If there are n_A elements in cluster A and n_B elements in cluster B, there are $n_A n_B$ such pairs. The single linkage (or nearest neighbor) method defines the distance between clusters as the minimum distance among all of these pairs. The complete linkage (or furthest neighbor) method defines the distance between clusters as the maximum distance among all of these pairs.

One of the more commonly used methods is Ward's method. At each stage, all potential mergers will reduce the number of current clusters by one. Each of these potential mergers will result in an increase in the overall within sum of squares. (The within sum of squares may be thought of as the amount of scatter about the group centroids. With n clusters the within sum of squares is equal to zero, since there is no scatter of other members about the group centroids. With one cluster, the within sum of squares is maximal.) Ward's method chooses that merger that results in the smallest increase in the within sum of squares. This is conceptually appealing, since we would like the within-group variability to remain as small as possible.

10.2.2 Illustration: 1990 Census Data for Erie County, New York

Here we will illustrate some of the features of cluster analysis using the dataset described above in the illustration of factor analysis.

Table 10.5 displays the results of a nonhierarchical k-means cluster analysis, where solutions range from $k = 2$ to $k = 4$. Three variables were used as clustering variables: the education variable, median household income, and the percentage of households headed by females. After standardization, the z-scores were used in the cluster analysis. For the two-cluster solution, the final cluster centers reveal that the first cluster is one where there are low scores on the education and median household income variables and high values on the percentage of households headed by females. The second cluster has the opposite characteristics, since the final cluster centroid is at education and income values that are above average, and at a location where the percentage of female-headed households is below average. There are 126 observations in the first cluster, and 105 in the second (and there are five observations with missing data). The ANOVA table reveals that all of the variables are contributing strongly to the success of the clustering, since all of the F-values are extremely high and significant. It is important to note that, since the cluster analysis is *designed* to make the F-statistic large by minimizing within-group variation, these F-statistics should not be interpreted in the usual way. In particular, we would *expect* the F-statistics to be large since we are creating clusters to make F large. Still, they can be used as rough guidelines to indicate

Table 10.5 **(a) Two-cluster solution; (b) three-cluster solution; (c) four-cluster solution**

(a)

Final Cluster Centers

	Cluster	
	1	2
Zscore(EDUC)	−.49757	.58646
Zscore(FEMALEH)	.51941	−.62329
Zscore(MEDHSINC)	−.55645	.76299

ANOVA

	Cluster		Error			
	Mean Square	df	Mean Square	df	F	Sig.
Zscore(EDUC)	67.302	1	.639	229	105.276	.000
Zscore(FEMALEH)	74.784	1	.678	229	110.335	.000
Zscore(MEDHSINC)	99.708	1	.496	229	201.103	.000

The F tests should be used only for descriptive purposes because the clusters have been chosen to maximize the differences among cases in different clusters. The observed significance levels are not corrected for this and thus cannot be interpreted as tests of the hypothesis that the cluster means are equal.

Number of Cases in each Cluster

Cluster	1	126.000
	2	105.000
Valid		231.000
Missing		5.000

(b)

Final Cluster Centers

	Cluster		
	1	2	3
Zscore(EDUC)	−.58897	−.27850	1.39280
Zscore(FEMALEH)	1.60968	−.27974	−.68224
Zscore(MEDHSINC)	−1.05910	.03641	1.11887

ANOVA

	Cluster		Error			
	Mean Square	df	Mean Square	df	F	Sig.
Zscore(EDUC)	57.715	2	.431	228	133.904	.000
Zscore(FEMALEH)	73.226	2	.366	228	199.830	.000
Zscore(MEDHSINC)	53.347	2	.467	228	114.152	.000

The F tests should be used only for descriptive purposes because the clusters have been chosen to maximize the differences among cases in different clusters. The observed significance levels are not corrected for this and thus cannot be interpreted as tests of the hypothesis that the cluster means are equal.

Number of Cases in each Cluster

Cluster	1	44.000
	2	141.000
	3	46.000
Valid		231.000
Missing		5.000

(c)

Final Cluster Centers

	Cluster			
	1	2	3	4
Zscore(EDUC)	4.76906	−.30937	.99199	−.62440
Zscore(FEMALEH)	−1.02424	−.17113	−.68459	1.98402
Zscore(MEDHSINC)	−.87318	−.11596	1.19929	−1.13847

ANOVA

	Cluster		Error			
	Mean Square	Df	Mean Square	df	F	Sig.
Zscore(EDUC)	41.594	3	.392	227	106.188	.000
Zscore(FEMALEH)	52.519	3	.319	227	164.566	.000
Zscore(MEDHSINC)	40.720	3	.401	227	101.478	.000

The F tests should be used only for descriptive purposes because the clusters have been chosen to maximize the differences among cases in diferent clusters. The observed significance levels are not corrected for this and thus cannot be interpreted as tests of the hypothesis that the cluster means are equal.

Number of Cases in each Cluster

Cluster	1	2.000
	2	143.000
	3	54.000
	4	32.000
Valid		231.000
Missing		5.000

the success of the clustering and the relative success that individual variables have in achieving the cluster solution.

The three-cluster solution is similar to the two-cluster solution, with the addition of a "middle" group that has values on all three variables that are close to the countywide averages. There are 41 observations in the first cluster (characterized by low levels of education and income and a high percentage of female-headed households), 141 observations in the middle group, and 46 tracts in the third group. Again, all of the *F*-statistics are high, implying that all three variables help to place the observations into clusters.

One of the groups in the four-cluster solution has only two observations. These two observations are characterized by an extremely high percentage of individuals with professional degrees.

It appears that there are two distinct clusters, with a third, fairly large group characterized by rather average values on the variables. In addition, the cluster analysis has been helpful in locating two census tracts that could be characterized as outliers due to their high values on the education variable. Figure 10.6 depicts the location of tracts in the three-cluster solution.

An important piece of output from a hierarchical cluster analysis is the *dendogram*. As its name implies, a dendogram is a tree-like diagram. It captures the history of the hierarchical clustering process as one proceeds from left to right along it. For illustrative purposes, it is rather difficult to show the dendogram that accompanies a hierarchical cluster analysis that has taken place for a

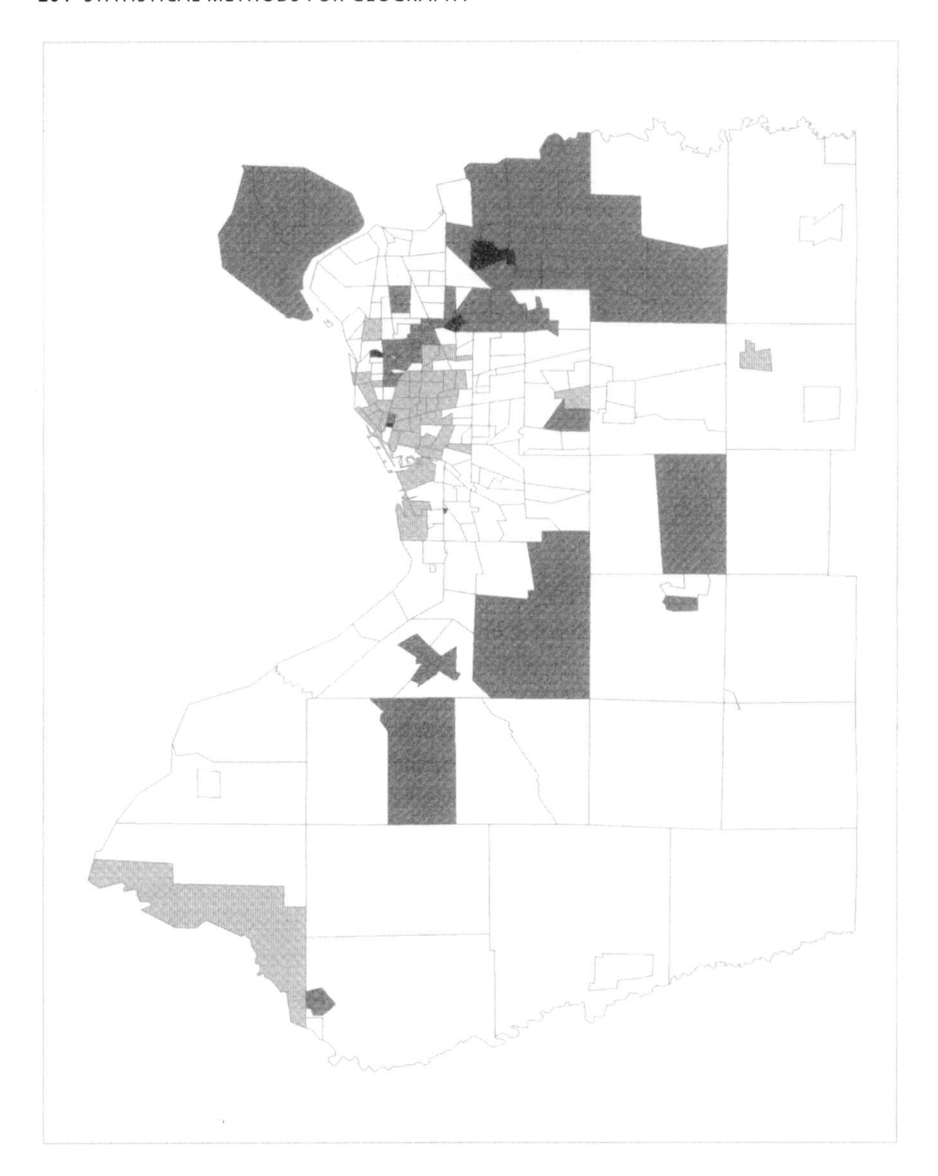

Figure 10.6 **Three-cluster solution**

very large number of observations. Instead, Figure 10.7 shows a dendogram for a subset of 30 tracts that have been selected at random from the dataset. At the left of the dendogram, the branches that meet indicate observations that have clustered together. For example, tracts 70 and 146 were very close together in the three-variable space and clustered together early in the process. In fact, the agglomeration schedule (shown in Table 10.6) indicates that these were the first two observations that were clustered together. The horizontal

* * * * * * H I E R A R C H I C A L C L U S T E R A N A L Y S I S * * * * * *

Dendrogram using Ward Method

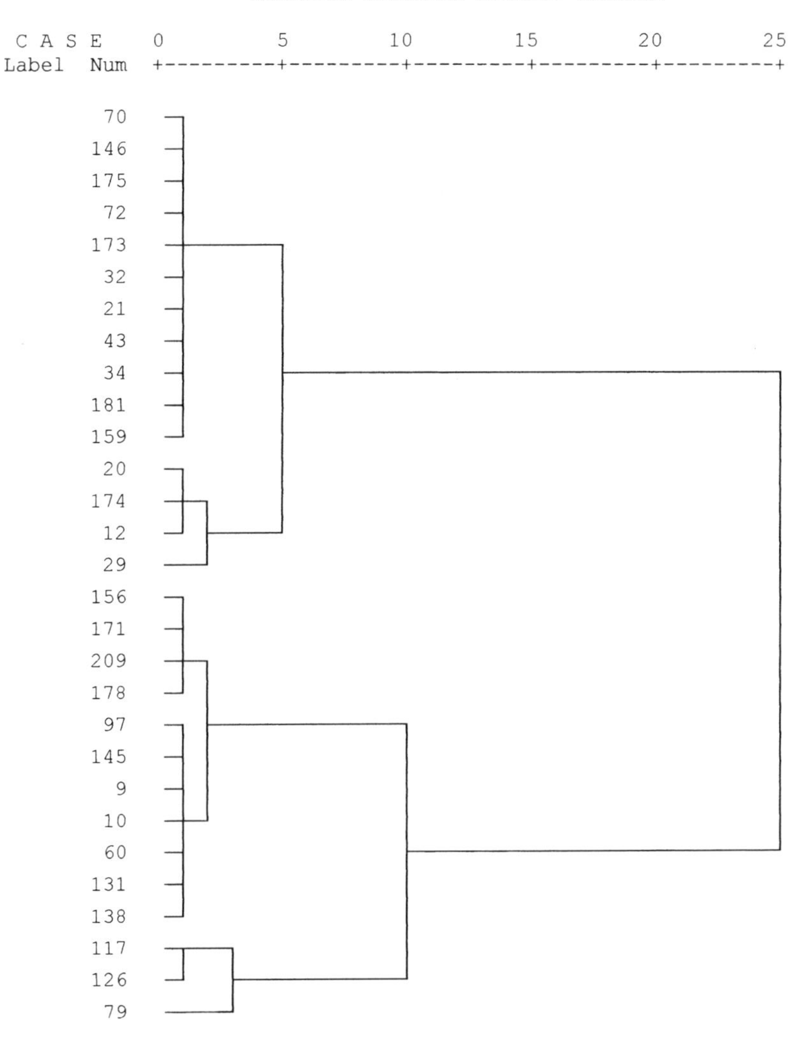

Figure 10.7 **Dendogram**

scale of the dendogram indicates the distance between the observations or groups that are clustered together. On the left of the dendogram, observations are close together when they cluster. On the right of the dendogram, there is only a small number of groups and the distance between those groups is larger.

Table 10.6 **Agglomeration schedule for hierarchical clustering**

Agglomeration Schedule

Stage	Cluster Combined		Coefficients	Stage Cluster First Appears		Next Stage
	Cluster 1	Cluster 2		Cluster 1	Cluster 2	
1	70	146	1.578E-02	0	0	7
2	34	181	3.158E-02	0	0	8
3	97	145	5.184E-02	0	0	9
4	72	173	7.689E-02	0	0	10
5	156	171	.104	0	0	14
6	21	43	.135	0	0	13
7	70	175	.168	1	0	17
8	34	159	.226	2	0	13
9	9	97	.295	0	3	12
10	32	72	.391	0	4	17
11	60	131	.514	0	0	19
12	9	10	.732	9	0	22
13	21	34	.978	6	8	21
14	156	209	1.237	5	0	20
15	20	174	1.545	0	0	16
16	12	20	1.858	0	15	23
17	32	70	2.259	10	7	21
18	117	126	2.721	0	0	25
19	60	138	3.193	11	0	22
20	156	178	3.782	14	0	24
21	21	32	4.642	13	17	26
22	9	60	5.718	12	19	24
23	12	29	7.126	16	0	26
24	9	156	8.771	22	20	27
25	79	117	11.396	0	18	27
26	12	21	17.329	23	21	28
27	9	79	28.934	24	25	28
28	9	12	58.941	27	26	0

To decide on the number of clusters, one can imagine taking a vertical line and proceeding from left to right along the dendogram. As one proceeds, the number of lines from the dendogram that intersect this vertical line decreases from n to 1. A good choice for the number of groups is one where there is a fairly large horizontal range in the dendogram where the number of groups does not change. In Figure 10.7, it would make little sense to choose five groups, since these five groups could easily be simplified into four by proceeding just a little further to the right on the dendogram. The figure shows that there are two clear groups of tracts. The tracts that are in each of these groups may be found by proceeding to the left from each of the two parallel, horizontal lines on the dendogram. Following all the way to the left, through all of the branches, reveals all of the tracts in each cluster. For example, one of the two clusters consists of observations 156, 171, 209, 178, 97, 145, 9, 10, 60, 131, 138, 117, 126, and 79. Note that a three-cluster solution would subdivide this particular cluster into two subclusters, and one of those subclusters would be quite small (consisting only of observations 117, 126, and 79). The next step in this analysis would be to examine the characteristics of the observations in each cluster. For example, observations 117, 126, and 79 all have quite high values on the education variable, coupled with high median household incomes.

10.3 Data Reduction Methods in *SPSS for Windows 9.0*

10.3.1 Factor Analysis

Click on Analyze, then Data Reduction, and then Factor. Choose the variables that will enter into the factor analysis. Under Rotation, choose Varimax (it is *not* the default). It is the most commonly used rotation method, and you should use it unless you have a good reason to choose an alternative! Under Extraction, it is most common to choose Principal Components and to choose as significant Eigenvalues over 1. These are the defaults, and so unless you wish to change them you do not have to do anything. Under Scores, choose Save as Variables. This will save the factor scores as new variables by attaching a number of columns to your dataset that is equal to the number of significant factors. Under Descriptive, choose Univariate Descriptives if desired. It is also useful to check the box labeled "coefficients" under Correlation Matrix, to print out a table of the correlation coefficients among variables.

10.3.2 Cluster Analysis

Hierarchical methods. Choose Analyze, then Classify, then Hierarchical Cluster. Next, choose the variables that are to be clustered. Then, under Method, choose the clustering method to be used. Note: Ward's method, though perhaps the most commonly used, is *not* the default choice. In fact, in Versions 8 and 9, one must scroll down the drop-down list to find it at the end of the list of methods. Next, choose the measure of distance that will be used; squared Euclidean distance is the default, and is a reasonable (and readily understandable!) choice. Still in this section, you will likely want to choose z-scores under the box labeled "standardize"; again, it is *not* the default option. Under Plots, one will often want to turn off the default "icicle plot" and check the box labeled "dendogram". Under Save, you may save cluster membership, which adds a column of data to the data table indicating the cluster to which each observation belongs. This can be done either for a single predefined choice of the number of clusters or for a predefined range of cluster numbers.

Nonhierarchical clustering. First click on Analyze, then on k-means. After choosing the variables to cluster (and recalling that it is usually a good idea to standardize the data by computing z-scores before doing this), choose the number of clusters desired. It is usually a good idea to click on Save, and save "cluster membership", which adds a column to the data table indicating the cluster membership of each observation.

Exercises

1. Explain and interpret the rotated factor loading table below

	Factor 1	Factor 2
% <15 yr old	.88	.21
% blue collar workers	.13	.86
% >65 yr old	−.92	−.11
% white collar workers	−.17	−.81
Median income	.24	−.71

2. Perform a hierarchical cluster analysis using the following data, and comment on the results.

Data for cluster analysis

Region	Mean age	%nonfamily	Median income (×000)
1	34	50	34
2	45	44	44
3	32	58	38
4	50	50	59
5	55	70	44
6	26	62	29
7	37	38	33
8	42	36	43
9	47	39	56
10	46	49	58
11	51	68	61
12	38	36	39
13	33	44	41
14	29	66	38

3. How many significant factors would be extracted in the following factor analysis? How many variables were in the original analysis? Explain your answer.

Factor	Eigenvalue	Cumulative percent of variance explained
1	3.0	45
2	2.5	70
3	1.5	78
4	0.9	89
5	0.3	92
6	0.2	94
7	0.2	96
8	0.2	98
9	0.1	99
10	0.1	100

4. A researcher collects the following information for a set of census tracts:

Tract	Median age	Income (×000)	% nonfamily	No. of autos	% new residents	% blue collar
1	26	29	32	1	23	33
2	35	38	24	2	21	21
3	48	49	29	3	16	44
4	47	55	55	3	18	44
5	36	39	66	2	23	41
6	29	32	42	2	33	40
7	55	58	38	3	10	31
8	56	66	36	3	11	24
9	29	32	33	1	23	28
10	33	44	29	2	21	29
11	44	49	31	2	18	31
12	47	46	38	2	15	30
13	51	52	55	3	12	20
14	44	49	52	2	18	19
15	37	40	38	1	19	43
16	38	41	34	2	21	31

Use factor analysis to summarize the data.
Assume that the data come from a 4×4 grid laid over the city as follows:

1	2	3	4
5	6	7	8
9	10	11	12
13	14	15	16

(a) Run a factor analysis to summarize the data above.
(b) How many factors are sufficient to describe the data (i.e., have eigenvalues greater than 1)?
(c) Describe the rotated factor loadings, describing each factor in terms of the most important variables that comprise it. Attempt to give names to the factors. (Note: in this part of the question, discuss only those factors with eigenvalues greater than 1.)
(d) Save the factor scores, and make a map of the scores on factor 1.

Epilogue

The primary purpose of this book has been to provide a foundation in some of the basic statistical tools that are used by geographers. The focus has been on inferential methods. Inferential statistical methods are attractive because they fit well into the time-honored framework of the scientific method. There are, of course, limitations to the use of these methods. Many of the concerns are related to the nature of hypothesis testing. Why do we test whether two populations have the same mean? An enumeration of two communities would almost certainly show that the true, "population" means of, say, commuting distance were in fact different. Why do we test whether a true regression coefficient is zero? Independent variables will almost always have *some* effect on the dependent variable, even if it is small. The principal point here is that in many situations the null hypothesis is not going to be true, so why are we testing it? A response to this concern is that the inferential framework provides not only a way of testing hypotheses, but also a way of establishing confidence intervals around estimated parameters. Thus we can state with a given level of confidence the magnitude of the difference in commuting times, and we can specify with a given level of precision the magnitude of a regression coefficient.

With the increasing availability of large datasets, there has been an appropriate development of exploratory methods. Such exploratory methods are extremely useful in "data mining" and "data trawling" to suggest new hypotheses. Ultimately, confirmation of these new hypotheses is called for, and inferential methods become more appropriate.

Where does the student of quantitative methods in geography go from here? The books by Longley *et al.* (1998) and Fotheringham *et al.* (2000) represent two good examples of recent attempts to summarize some of the new developments in the field. They include accounts of new developments in the areas of exploratory spatial data analysis and geocomputation. In order of increasing mathematical sophistication are books on spatial data analysis by Bailey and Gatrell (1995), Haining (1990a), and Cressie (1993). These require more background than that required for this book, but that should not deter the interested student from exploring them. Even if one does not absorb all of the mathematical detail, it is possible to get a good sense of the types of questions and the range of problems that spatial analysis can address.

Selected Publications

The following is a short list of some recent examples of the use of statistical methods in geography. Full details can be found in the Reference list.

Two-sample tests
  Nelson 1997
Factor analysis in regression
  Ormrod and Cole 1996
  O'Reilly and Webster 1998
  Ackerman 1998
Logistic regression
  Myers *et al.* 1997
Correlation
  Williams and Parker 1997
  Allen and Turner 1996
Spearman's
  Keim 1997
Correlation and regression
  Wyllie and Smith 1996 (stepwise regression)
Regression
  Fan 1996
Principal components and factor analysis
  Clarke and Holly 1996
  Webster 1996
Cluster analysis
  Comrie 1996
  Dagel 1997

Appendix A: Statistical Tables

Table A.1 **Random digits**

Each location in the table is equally likely to be the digit 0, 1, 2, 3, 4, 5, 6, 7, 8, or 9.

39203	59841	91168	32021	82081	60164	37385	52925	91004	71887
39965	79079	97829	95836	26651	12495	68275	20281	73978	07258
17752	87652	07004	95860	89325	56997	70904	91993	13209	50274
04284	63927	07533	60557	41339	16728	96512	11116	92345	04612
03440	97786	37416	24541	36408	63936	36480	87028	05094	95318
07466	12899	31434	06525	81175	38234	24468	30891	89620	50129
83343	72721	52695	36309	67961	73792	63300	89222	10618	24229
03745	48015	85373	77206	76214	85412	83510	73998	13500	65084
27975	70407	56983	07913	38682	89173	40739	40168	95705	46872
54284	28109	48080	80215	85753	64411	27938	56201	16005	49409
79521	93795	56291	03839	16098	44436	22678	37566	45822	26879
17817	48797	59971	28104	68171	05068	98190	33721	13991	73487
56213	82716	77356	91791	31267	19598	25159	28785	57736	72346
75194	03658	65212	50828	73031	12498	30153	80522	30866	05307
44549	28479	49939	43539	66337	61547	25104	27361	27060	17720
11543	45735	21121	46119	96548	48237	30815	01082	00715	18213
27327	47369	72686	74153	67849	91820	22255	91564	28009	19796
65332	83444	40231	84229	48713	46748	54693	63440	03439	97497
45214	30409	35466	73494	39421	86061	88928	55676	68453	66827
77929	36175	61017	71350	93393	32687	29040	74575	45306	22552
54366	88887	16301	19105	51147	31217	41907	42982	64904	63597
08535	65466	48869	58315	23905	24696	66332	22822	37808	78375
36947	67802	81864	59051	52076	34284	06530	51015	39540	61780
28323	33789	56413	16652	28571	53781	63579	42659	53203	29708
16748	41349	75175	66405	75745	33003	32043	01747	49361	61584
33178	69744	11252	49458	86585	85536	92257	24864	48761	31924
26466	93243	88962	31547	05650	29480	92795	39219	22342	60169
36535	14197	72029	40094	61100	17633	38541	08250	04353	13417
66835	93340	09121	97179	24446	47809	87930	83677	46036	07924
09357	02826	35480	92998	35244	39454	50956	36244	31511	40640
07296	75285	29833	78926	48012	97299	56635	57142	00203	77302
01106	48819	40679	96311	90666	91712	16907	65802	94408	76429
15742	99837	87999	36431	96530	84598	62879	82602	57911	18505
16523	51356	37907	65491	39889	49415	97503	09430	39471	12136
03536	42548	50478	54022	18614	03129	68513	08643	91870	93123
73445	35057	97928	83183	57729	35701	70757	28092	97686	90810
52017	99654	63051	87131	87755	29329	52001	24808	54075	48002
63724	57039	06679	46472	92762	75952	54470	88720	57702	61299
16675	01990	38803	84706	24066	41937	26551	58381	04810	35915
01377	36919	49327	24518	61098	25962	04427	33234	04480	02438
49752	61849	05823	84198	18174	74419	10322	95196	47893	77825
40734	81595	96763	68282	34155	29452	94005	23972	66115	40478
64213	91973	62604	00789	21825	25568	00981	89250	24446	86013
24505	41214	03031	34756	31600	84374	36871	83645	80482	22081
34248	31337	78109	49077	10187	84757	45754	51435	52726	24296
60229	06451	61294	53777	17640	85533	10178	23212	02002	08264
36712	16560	35055	99750	53169	58659	37377	53580	16829	10472
94150	42762	54989	58564	12434	81297	36197	84099	55629	03717
36402	94992	51794	59245	87178	84460	58370	34416	75064	07568
15853	95261	90876	66395	72788	66605	08718	96740	45414	81015

Table A.1 **(Continued)**

84807	71928	78331	51465	39259	63729	32989	80330	57238	98955
98408	62427	04782	69732	83461	01420	68618	11575	24972	14040
61825	69602	11652	56412	22210	03517	40796	29470	49044	10343
39883	29540	45090	05811	62559	50967	66031	48501	05426	82446
68403	57420	50632	05400	81552	91661	37190	95155	26634	01135
58917	60176	48503	14559	18274	45809	09748	19716	15081	84704
72565	19292	16976	41309	04164	94000	19939	55374	26109	58722
58272	12730	89732	49176	14281	57181	02887	84072	91832	97489
92754	47117	98296	74972	38940	45352	58711	43014	95376	57402
34520	96779	25092	96327	05785	76439	10332	07534	79067	27126
18388	17135	08468	31149	82568	96509	32335	65895	64362	01431
06578	34257	67618	62744	93422	89236	53124	85750	98015	00038
67183	75783	54437	58890	02256	53920	61369	65913	65478	62319
26942	92564	92010	95670	75547	20940	06219	28040	10050	05974
06345	01152	49596	02064	85321	59627	28489	88186	74006	18320
24221	12108	16037	99857	73773	42506	60530	96317	29918	16918
83975	61251	82471	06941	48817	76078	68930	39693	87372	09600
86232	01398	50258	22868	71052	10127	48729	67613	59400	65886
04912	01051	33687	03296	17112	23843	16796	22332	91570	47197
15455	88237	91026	36454	18765	97891	11022	98774	00321	10386
88430	09861	45098	66176	59598	98527	11059	31626	10798	50313
48849	11583	63654	55670	89474	75232	14186	52377	19129	67166
33659	59617	40920	30295	07463	79923	83393	77120	38862	75503
60198	41729	19897	04805	09351	76734	10333	87776	36947	88618
55868	53145	66232	52007	81206	89543	66226	45709	37114	78075
22011	71396	95174	43043	68304	56773	83931	43631	50995	68130
90301	54934	08008	00565	67790	84760	82229	64147	28031	11609
07586	90936	21021	54066	87281	63574	41155	01740	29025	19909
09973	76136	87904	54419	34370	75071	56201	16768	61934	12083
59750	42528	19864	31595	72097	17005	24682	43560	74423	59197
74492	19327	17812	63897	65708	07709	13817	95943	07909	75504
69042	57646	38606	30549	34351	21432	50312	10566	43842	70046
16054	32268	29828	73413	53819	39324	13581	71841	94894	64223
17930	78622	70578	23048	73730	73507	69602	77174	32593	45565
46812	93896	65639	73905	45396	71653	01490	33674	16888	53434
04590	07459	04096	15216	56633	69845	85550	15141	56349	56117
99618	63788	86396	37564	12962	96090	70358	23378	63441	36828
34545	32273	45427	30693	49369	27427	28362	17307	45092	08302
04337	00565	27718	67942	19284	69126	51649	03469	88009	41916
73810	70135	72055	90111	71202	08210	76424	66364	63081	37784
60555	94102	39146	67795	05985	43280	97202	35613	25369	47959
58261	16861	39080	22820	46555	32213	38440	32662	48259	61197
98765	65802	44467	03358	38894	34290	31107	25519	26585	34852
39157	58231	30710	09394	04012	49122	26283	34946	23590	25663
08143	91252	23181	51183	52102	85298	52008	48688	86779	21722
66806	72352	64500	89120	13493	85813	93999	12558	24852	04575
08289	82806	36490	96421	81718	63075	54178	39209	03050	47089
12989	31280	71466	72234	26922	04753	61943	86149	26938	53736
44154	63471	30657	62298	56461	48879	54108	97126	43219	95349
63788	18000	10049	49041	28807	64190	39753	17397	48026	76947

Table A.2 **Normal distribution**

The tabled entries represent the proportion p of area under the normal curve above the indicated values of z. (Example: .0694 or 6.94% of the area is above $z = 1.48$). For negative values of z, the tabled entries represent the area less than $-z$. (Example: .3015 or 30.15% of the area is beneath $z = -.52$.)

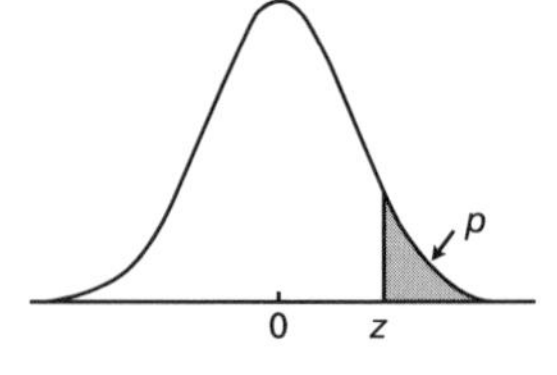

	Second decimal place of z									
z	.00	.01	.02	.03	.04	.05	.06	.07	.08	.09
0.0	.5000	.4960	.4920	.4880	.4840	.4801	.4761	.4721	.4681	.4641
0.1	.4602	.4562	.4522	.4483	.4443	.4404	.4364	.4325	.4286	.4247
0.2	.4207	.4168	.4129	.4090	.4052	.4013	.3974	.3936	.3897	.3859
0.3	.3821	.3783	.3745	.3707	.3669	.3632	.3594	.3557	.3520	.3483
0.4	.3446	.3409	.3372	.3336	.3300	.3264	.3228	.3192	.3156	.3121
0.5	.3085	.3050	.3015	.2981	.2946	.2912	.2877	.2843	.2810	.2776
0.6	.2743	.2709	.2676	.2643	.2611	.2578	.2546	.2514	.2483	.2451
0.7	.2420	.2389	.2358	.2327	.2297	.2266	.2236	.2206	.2177	.2148
0.8	.2119	.2090	.2061	.2033	.2005	.1977	.1949	.1922	.1894	.1867
0.9	.1841	.1814	.1788	.1762	.1736	.1711	.1685	.1660	.1635	.1611
1.0	.1587	.1562	.1539	.1515	.1492	.1469	.1446	.1423	.1401	.1379
1.1	.1357	.1335	.1314	.1292	.1271	.1251	.1230	.1210	.1190	.1170
1.2	.1151	.1131	.1112	.1093	.1075	.1056	.1038	.1020	.1003	.0985
1.3	.0968	.0951	.0934	.0918	.0901	.0885	.0869	.0853	.0838	.0823
1.4	.0808	.0793	.0778	.0764	.0749	.0735	.0721	.0708	.0694	.0681
1.5	.0668	.0655	.0643	.0630	.0618	.0606	.0594	.0582	.0571	.0559
1.6	.0548	.0537	.0526	.0516	.0505	.0495	.0485	.0475	.0465	.0455
1.7	.0446	.0436	.0427	.0418	.0409	.0401	.0392	.0384	.0375	.0367
1.8	.0359	.0351	.0344	.0336	.0329	.0322	.0314	.0307	.0301	.0294
1.9	.0287	.0281	.0274	.0268	.0262	.0256	.0250	.0244	.0239	.0233
2.0	.0228	.0222	.0217	.0212	.0207	.0202	.0197	.0192	.0188	.0183
2.1	.0179	.0174	.0170	.0166	.0162	.0158	.0154	.0150	.0146	.0143
2.2	.0139	.0136	.0132	.0129	.0125	.0122	.0119	.0116	.0113	.0110
2.3	.0107	.0104	.0102	.0099	.0096	.0094	.0091	.0089	.0087	.0084
2.4	.0082	.0080	.0078	.0075	.0073	.0071	.0069	.0068	.0066	.0064
2.5	.0062	.0060	.0059	.0057	.0055	.0054	.0052	.0051	.0049	.0048
2.6	.0047	.0045	.0044	.0043	.0041	.0040	.0039	.0038	.0037	.0036
2.7	.0035	.0034	.0033	.0032	.0031	.0030	.0029	.0028	.0027	.0026
2.8	.0026	.0025	.0024	.0023	.0023	.0022	.0021	.0021	.0020	.0019
2.9	.0019	.0018	.0018	.0017	.0016	.0016	.0015	.0015	.0014	.0014
3.0	.0013	.0013	.0013	.0012	.0012	.0011	.0011	.0011	.0010	.0010

Adapted with rounding from Table II of Fisher and Yates 1974.

Table A.3 **Student's *t* distribution**

For various degrees of freedom (df), the tabled entries represent the critical values of t above which a specified proportion p of the t distribution falls. (Example: for df = 9, a t of 2.262 is surpassed by .025 or 2.5% of the total distribution.

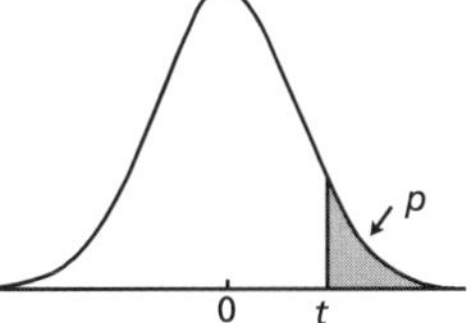

	p (one-tailed probabilities)				
df	.10	.05	.025	.01	.005
1	3.078	6.314	12.706	31.821	63.657
2	1.886	2.920	4.303	6.965	9.925
3	1.638	2.353	3.182	4.541	5.841
4	1.533	2.132	2.776	3.747	4.604
5	1.476	2.015	2.571	3.365	4.032
6	1.440	1.943	2.447	3.143	3.707
7	1.415	1.895	2.365	2.998	3.499
8	1.397	1.860	2.306	2.896	3.355
9	1.383	1.833	2.262	2.821	3.250
10	1.372	1.812	2.228	2.764	3.169
11	1.363	1.796	2.201	2.718	3.106
12	1.356	1.782	2.179	2.681	3.055
13	1.350	1.771	2.160	2.650	3.012
14	1.345	1.761	2.145	2.624	2.977
15	1.341	1.753	2.131	2.602	2.947
16	1.337	1.746	2.120	2.583	2.921
17	1.333	1.740	2.110	2.567	2.898
18	1.330	1.734	2.101	2.552	2.878
19	1.328	1.729	2.093	2.539	2.861
20	1.325	1.725	2.086	2.528	2.845
21	1.323	1.721	2.080	2.518	2.831
22	1.321	1.717	2.074	2.508	2.819
23	1.319	1.714	2.069	2.500	2.807
24	1.318	1.711	2.064	2.492	2.797
25	1.316	1.708	2.060	2.485	2.787
26	1.315	1.706	2.056	2.479	2.779
27	1.314	1.703	2.052	2.473	2.771
28	1.313	1.701	2.048	2.467	2.763
29	1.311	1.699	2.045	2.462	2.756
30	1.310	1.697	2.042	2.457	2.750
40	1.303	1.684	2.021	2.423	2.704
60	1.296	1.671	2.000	2.390	2.660
120	1.289	1.658	1.980	2.358	2.617
∞	1.282	1.645	1.960	2.326	2.576

Adapted from Table III of Fisher and Yates (1974).

Table A.4 **Cumulative distribution of Student's *t* distribution**

t\v	1	2	3	4	5	6	7	8	9	10
0.0	0.50000	0.50000	0.50000	0.50000	0.50000	0.50000	0.50000	0.50000	0.50000	0.50000
0.1	0.53173	0.53527	0.53667	0.53742	0.53788	0.53820	0.53843	0.53860	0.53873	0.53884
0.2	0.56283	0.57002	0.57286	0.57438	0.57532	0.57596	0.57642	0.57676	0.57704	0.57726
0.3	0.56283	0.57002	0.57286	0.57438	0.57532	0.57596	0.57642	0.57676	0.57704	0.57726
0.4	0.62112	0.63608	0.64203	0.64520	0.64716	0.64850	0.64946	0.65019	0.65076	0.65122
0.5	0.64758	0.66667	0.67428	0.67834	0.68085	0.68256	0.68380	0.68473	0.68546	0.68605
0.6	0.67202	0.69529	0.70460	0.70958	0.71267	0.71477	0.71629	0.71745	0.71835	0.71907
0.7	0.69440	0.72181	0.73284	0.73875	0.74243	0.74493	0.74674	0.74811	0.74919	0.75006
0.8	0.71478	0.74618	0.75890	0.76574	0.76999	0.77289	0.77500	0.77659	0.77784	0.77885
0.9	0.73326	0.76845	0.78277	0.79050	0.79531	0.79860	0.80099	0.80280	0.80422	0.80536
1.0	0.75000	0.78868	0.80450	0.81305	0.81839	0.82204	0.82469	0.82670	0.82828	0.82955
1.1	0.76515	0.80698	0.82416	0.83346	0.83927	0.84325	0.84614	0.84834	0.85006	0.85145
1.2	0.77886	0.82349	0.84187	0.85182	0.85805	0.86232	0.86541	0.86777	0.86961	0.87110
1.3	0.79129	0.83838	0.85777	0.86827	0.87485	0.87935	0.88262	0.88510	0.88705	0.88862
1.4	0.80257	0.85177	0.87200	0.88295	0.88980	0.89448	0.89788	0.90046	0.90249	0.90412
1.5	0.81283	0.86380	0.88471	0.89600	0.90305	0.90786	0.91135	0.91400	0.91608	0.91775
1.6	0.82219	0.87464	0.89605	0.90758	0.91475	0.91964	0.92318	0.92587	0.92797	0.92966
1.7	0.83075	0.88439	0.90615	0.91782	0.92506	0.92998	0.93354	0.93622	0.93833	0.94002
1.8	0.83859	0.89317	0.91516	0.92688	0.93412	0.93902	0.94256	0.94522	0.94731	0.94897
1.9	0.84579	0.90109	0.92318	0.93488	0.94207	0.94691	0.95040	0.95302	0.95506	0.95669
2.0	0.85242	0.90825	0.93034	0.94194	0.94903	0.95379	0.95719	0.95974	0.96172	0.96331
2.1	0.85854	0.91473	0.93672	0.94817	0.95512	0.95976	0.96306	0.96553	0.96744	0.96896
2.2	0.86420	0.92060	0.94241	0.95367	0.96045	0.96495	0.96813	0.97050	0.97233	0.97378
2.3	0.86945	0.92593	0.94751	0.95853	0.96511	0.96945	0.97250	0.97476	0.97650	0.97787
2.4	0.87433	0.93077	0.95206	0.96282	0.96919	0.97335	0.97627	0.97841	0.98005	0.98134
2.5	0.87888	0.93519	0.95615	0.96662	0.97275	0.97674	0.97950	0.98153	0.98307	0.98428
2.6	0.88313	0.93923	0.95981	0.96998	0.97587	0.97967	0.98229	0.98419	0.98563	0.98675
2.7	0.88709	0.94292	0.96311	0.97295	0.97861	0.98221	0.98468	0.98646	0.98780	0.98884
2.8	0.89081	0.94630	0.96607	0.97559	0.98100	0.98442	0.98674	0.98840	0.98964	0.99060
2.9	0.89430	0.94941	0.96875	0.97794	0.98310	0.98633	0.98851	0.99005	0.99120	0.99208
3.0	0.89758	0.95227	0.97116	0.98003	0.98495	0.98800	0.99003	0.99146	0.99252	0.99333
3.1	0.90067	0.95490	0.97335	0.98189	0.98657	0.98944	0.99134	0.99267	0.99364	0.99437
3.2	0.90359	0.95733	0.97533	0.98355	0.98800	0.99070	0.99247	0.99369	0.99459	0.99525
3.3	0.90634	0.95958	0.97713	0.98503	0.98926	0.99180	0.99344	0.99457	0.99539	0.99599
3.4	0.90895	0.96166	0.97877	0.98636	0.99037	0.99275	0.99428	0.99532	0.99606	0.99661
3.5	0.91141	0.96358	0.98026	0.98755	0.99136	0.99359	0.99500	0.99596	0.99664	0.99714
3.6	0.91376	0.96538	0.98162	0.98862	0.99223	0.99432	0.99563	0.99651	0.99713	0.99758
3.7	0.91598	0.96705	0.98286	0.98958	0.99300	0.99496	0.99617	0.99698	0.99754	0.99795
3.8	0.91809	0.96860	0.98400	0.99045	0.99369	0.99552	0.99664	0.99738	0.99789	0.99826
3.9	0.92010	0.97005	0.98504	0.99123	0.99430	0.99601	0.99705	0.99773	0.99819	0.99852
4.0	0.92202	0.97141	0.98600	0.99193	0.99484	0.99644	0.99741	0.99803	0.99845	0.99874
4.2	0.92560	0.97386	0.98768	0.99315	0.99575	0.99716	0.99798	0.99850	0.99885	0.99909
4.4	0.92887	0.97602	0.98912	0.99415	0.99649	0.99772	0.99842	0.99886	0.99914	0.99933
4.6	0.93186	0.97792	0.99034	0.99498	0.99708	0.99815	0.99876	0.99912	0.99936	0.99951
4.8	0.93462	0.97962	0.99140	0.99568	0.99756	0.99850	0.99902	0.99932	0.99951	0.99964
5.0	0.93717	0.98113	0.99230	0.99625	0.99795	0.99877	0.99922	0.99947	0.99963	0.99973
5.2	0.93952	0.98248	0.99309	0.99674	0.99827	0.99899	0.99937	0.99959	0.99972	0.99980
5.4	0.94171	0.98369	0.99378	0.99715	0.99853	0.99917	0.99950	0.99968	0.99978	0.99985
5.6	0.94375	0.98478	0.99437	0.99750	0.99875	0.99931	0.99959	0.99975	0.99983	0.99989
5.8	0.94565	0.98577	0.99490	0.99780	0.99893	0.99942	0.99967	0.99980	0.99987	0.99991
6.0	0.94743	0.98666	0.99536	0.99806	0.99908	0.99952	0.99973	0.99984	0.99990	0.99993
6.2	0.94910	0.98748	0.99577	0.99828	0.99920	0.99959	0.99978	0.99987	0.99992	0.99995
6.4	0.95066	0.98822	0.99614	0.99847	0.99931	0.99966	0.99982	0.99990	0.99994	0.99996
6.6	0.95214	0.98890	0.99646	0.99863	0.99940	0.99971	0.99985	0.99992	0.99995	0.99997
6.8	0.95352	0.98953	0.99675	0.99878	0.99948	0.99975	0.99987	0.99993	0.99996	0.99998
7.0	0.95483	0.99010	0.99701	0.99890	0.99954	0.99979	0.99990	0.99994	0.99997	0.99998
7.2	0.95607	0.99063	0.99724	0.99901	0.99960	0.99982	0.99991	0.99995	0.99997	0.99999
7.4	0.95724	0.99111	0.99745	0.99911	0.99964	0.99984	0.99993	0.99996	0.99998	0.99999
7.6	0.95836	0.99156	0.99764	0.99920	0.99969	0.99986	0.99994	0.99997	0.99998	0.99999
7.8	0.95941	0.99198	0.99781	0.99927	0.99972	0.99988	0.99995	0.99997	0.99999	0.99999
8.0	0.96042	0.99237	0.99796	0.99934	0.99975	0.99990	0.99996	0.99998	0.99999	0.99999

Source: E.S. Pearson and H.O. Hartley (eds.) [1966], *Biometrika Tables for Statisticians*, vol. 1, Cambridge University Press, Cambridge, England (by permission).

Table A.4 **(Continued)**

t\v	11	12	13	14	15	16	17	18	19	20
0.0	0.50000	0.50000	0.50000	0.50000	0.50000	0.50000	0.50000	0.50000	0.50000	0.50000
0.1	0.53893	0.53900	0.53907	0.53912	0.53917	0.53921	0.53924	0.53928	0.53930	0.53933
0.2	0.57744	0.57759	0.57771	0.57782	0.57792	0.57800	0.57807	0.57814	0.57820	0.57825
0.3	0.61511	0.61534	0.61554	0.61571	0.61585	0.61598	0.61609	0.61619	0.61628	0.61636
0.4	0.65159	0.65191	0.65217	0.65240	0.65260	0.65278	0.65293	0.65307	0.65319	0.65330
0.5	0.68654	0.68694	0.68728	0.68758	0.68783	0.68806	0.68826	0.68843	0.68859	0.68873
0.6	0.71967	0.72017	0.72059	0.72095	0.72127	0.72155	0.72179	0.72201	0.72220	0.72238
0.7	0.75077	0.75136	0.75187	0.75230	0.75268	0.75301	0.75330	0.75356	0.75380	0.75400
0.8	0.77968	0.78037	0.78096	0.78146	0.78190	0.78229	0.78263	0.78293	0.78320	0.78344
0.9	0.80630	0.80709	0.80776	0.80883	0.80883	0.80927	0.80965	0.81000	0.81031	0.81058
1.0	0.83060	0.83148	0.83222	0.83286	0.83341	0.83390	0.83433	0.83472	0.83506	0.83537
1.1	0.85259	0.85355	0.85436	0.85506	0.85566	0.85620	0.85667	0.85709	0.85746	0.85780
1.2	0.87233	0.87335	0.87422	0.87497	0.87562	0.87620	0.87670	0.87715	0.87756	0.87792
1.3	0.88991	0.89099	0.89191	0.89270	0.89339	0.98399	0.89452	0.89500	0.89542	0.89581
1.4	0.90546	0.90658	0.90754	0.90836	0.90907	0.90970	0.91025	0.91074	0.91118	0.91158
1.5	0.91912	0.92027	0.92125	0.92209	0.92282	0.92346	0.92402	0.92452	0.92498	0.92538
1.6	0.93105	0.93221	0.93320	0.93404	0.93478	0.93542	0.93599	0.93650	0.93695	0.93736
1.7	0.94140	0.94256	0.94354	0.94439	0.94512	0.94576	0.94632	0.94683	0.94728	0.94768
1.8	0.95034	0.95148	0.95245	0.95328	0.95400	0.95463	0.95518	0.95568	0.95612	0.95652
1.9	0.95802	0.95914	0.96008	0.96089	0.96158	0.96220	0.96273	0.96321	0.96364	0.96403
2.0	0.96460	0.96567	0.96658	0.96736	0.96803	0.96861	0.96913	0.96959	0.97000	0.97037
2.1	0.97020	0.97123	0.97209	0.97283	0.97347	0.97403	0.97452	0.97495	0.97534	0.97569
2.2	0.97496	0.97593	0.97675	0.97745	0.97805	0.97858	0.97904	0.97945	0.97981	0.98014
2.3	0.86945	0.92593	0.94751	0.95853	0.96511	0.96945	0.97250	0.97476	0.97650	0.97787
2.4	0.98238	0.98324	0.98396	0.98457	0.98509	0.98554	0.98594	0.98629	0.98660	0.98688
2.5	0.98525	0.98604	0.98671	0.98727	0.98775	0.98816	0.98853	0.98885	0.98913	0.98938
2.6	0.98765	0.98839	0.98900	0.98951	0.98995	0.99033	0.99066	0.99095	0.99121	0.99144
2.7	0.98967	0.99035	0.99090	0.99137	0.99177	0.99211	0.99241	0.99267	0.99290	0.99311
2.8	0.99136	0.99198	0.99249	0.99291	0.99327	0.99358	0.99385	0.99408	0.99429	0.99447
2.9	0.99278	0.99334	0.99380	0.99418	0.99450	0.99478	0.99502	0.99523	0.99541	0.99557
3.0	0.99396	0.99447	0.99488	0.99522	0.99551	0.99576	0.99597	0.99616	0.99632	0.99646
3.1	0.99495	0.99541	0.99578	0.99608	0.99634	0.99656	0.99675	0.99691	0.99705	0.99718
3.2	0.99577	0.99618	0.99652	0.99679	0.99702	0.99721	0.99738	0.99752	0.99764	0.99775
3.3	0.99646	0.99683	0.99713	0.99737	0.99757	0.99774	0.99789	0.99801	0.99812	0.99821
3.4	0.99703	0.99737	0.99763	0.99784	0.99802	0.99817	0.99830	0.99840	0.99850	0.99858
3.5	0.99751	0.99781	0.99804	0.99823	0.99839	0.99852	0.99863	0.99872	0.99880	0.99887
3.6	0.99791	0.99818	0.99838	0.99855	0.99869	0.99880	0.99890	0.99898	0.99905	0.99911
3.7	0.99825	0.99848	0.99867	0.99881	0.99893	0.99903	0.99911	0.99918	0.99924	0.99929
3.8	0.99853	0.99874	0.99890	0.99902	0.99913	0.99921	0.99928	0.99934	0.99939	0.99944
3.9	0.99876	0.99895	0.99909	0.99920	0.99929	0.99936	0.99942	0.99948	0.99952	0.99956
4.0	0.99896	0.99912	0.99924	0.99934	0.99942	0.99948	0.99954	0.99958	0.99962	0.99965
4.2	0.99926	0.99938	0.99948	0.99955	0.99961	0.99966	0.99970	0.99973	0.99976	0.99978
4.4	0.99947	0.99957	0.99964	0.99970	0.99974	0.99978	0.99980	0.99983	0.99985	0.99986
4.6	0.99962	0.99969	0.99975	0.99979	0.99983	0.99985	0.99987	0.99989	0.99990	0.99991
4.8	0.99972	0.99978	0.99983	0.99986	0.99988	0.99990	0.99992	0.99993	0.99994	0.99995
5.0	0.99980	0.99985	0.99988	0.99990	0.99992	0.99993	0.99995	0.99995	0.99996	0.99997
5.2	0.99985	0.99989	0.99992	0.99993	0.99995	0.99996	0.99996	0.99997	0.99997	0.99998
5.4	0.99989	0.99992	0.99994	0.99995	0.99996	0.99997	0.99998	0.99998	0.99998	0.99999
5.6	0.99992	0.99994	0.99996	0.99997	0.99997	0.99998	0.99998	0.99999	0.99999	0.99999
5.8	0.99994	0.99996	0.99997	0.99998	0.99998	0.99999	0.99999	0.99999	0.99999	0.99999
6.0	0.99995	0.99997	0.99998	0.99998	0.99999	0.99999	0.99999	0.99999		
6.2	0.99997	0.99998	0.99998	0.99999	0.99999	0.99999				
6.4	0.99997	0.99998	0.99999	0.99999	0.99999					
6.6	0.99998	0.99999	0.99999	0.99999						
6.8	0.99998	0.99999	0.99999							
7.00	0.99999	0.99999								

Table A.5 *F* distribution

For various pairs of degrees of freedom υ_1, υ_2, the tabled entries represent the critical values of *F* above which a proportion *p* of the distribution falls. (Example: for df = 4,16 an $F = 2.33$ is exceeded by $p >= .10$ of the distribution.) Tables are provided for values of *p* equal to .10, .05, .01.

$p = .10$ values

Degrees of freedom for denominator υ_2	Degrees of freedom for numerator, υ_1: 1	2	3	4	5	6	7	8	9	10	12	15	20	30	40	60	120	∞
1	39.86	49.50	53.59	55.83	57.24	58.20	58.91	59.44	59.86	60.19	60.71	61.22	61.74	62.26	62.53	62.79	63.06	63.33
2	8.53	9.00	9.16	9.24	9.29	9.33	9.35	9.37	9.38	9.39	9.41	9.42	9.44	9.46	9.47	9.47	9.48	9.49
3	5.54	5.46	5.39	5.34	5.31	5.28	5.27	5.25	5.24	5.23	5.22	5.20	5.18	5.17	5.16	5.15	5.14	5.13
4	4.54	4.32	4.19	4.11	4.05	4.01	3.98	3.95	3.94	3.92	3.90	3.87	3.84	3.82	3.80	3.79	3.78	3.76
5	4.06	3.78	3.62	3.52	3.45	3.40	3.37	3.34	3.32	3.30	3.27	3.24	3.21	3.17	3.16	3.14	3.12	3.10
6	3.78	3.46	3.29	3.18	3.11	3.05	3.01	2.98	2.96	2.94	2.90	2.87	2.84	2.80	2.78	2.76	2.74	2.72
7	3.59	3.26	3.07	2.96	2.88	2.83	2.78	2.75	2.72	2.70	2.67	2.63	2.59	2.56	2.54	2.51	2.49	2.47
8	3.46	3.11	2.92	2.81	2.73	2.67	2.62	2.59	2.56	2.54	2.50	2.46	2.42	2.38	2.36	2.34	2.32	2.29
9	3.36	3.01	2.81	2.69	2.61	2.55	2.51	2.47	2.44	2.42	2.38	2.34	2.30	2.25	2.23	2.21	2.18	2.16
10	3.29	2.92	2.73	2.61	2.52	2.46	2.41	2.38	2.35	2.32	2.28	2.24	2.20	2.16	2.13	2.11	2.08	2.06
11	3.23	2.86	2.66	2.54	2.45	2.39	2.34	2.30	2.27	2.25	2.21	2.17	2.12	2.08	2.05	2.03	2.00	1.97
12	3.18	2.81	2.61	2.48	2.39	2.33	2.28	2.24	2.21	2.19	2.15	2.10	2.06	2.01	1.99	1.96	1.93	1.90
13	3.14	2.76	2.56	2.43	2.35	2.28	2.23	2.20	2.16	2.14	2.10	2.05	2.01	1.96	1.93	1.90	1.88	1.85
14	3.10	2.73	2.52	2.39	2.31	2.24	2.19	2.15	2.12	2.10	2.05	2.01	1.96	1.91	1.89	1.86	1.83	1.80
15	3.07	2.70	2.49	2.36	2.27	2.21	2.16	2.12	2.09	2.06	2.02	1.97	1.92	1.87	1.85	1.82	1.79	1.76
16	3.05	2.67	2.46	2.33	2.24	2.18	2.13	2.09	2.06	2.03	1.99	1.94	1.89	1.84	1.81	1.78	1.75	1.72
17	3.03	2.64	2.44	2.31	2.22	2.15	2.10	2.06	2.03	2.00	1.96	1.91	1.86	1.81	1.78	1.75	1.72	1.69
18	3.01	2.62	2.42	2.29	2.20	2.13	2.08	2.04	2.00	1.98	1.93	1.89	1.84	1.78	1.75	1.72	1.69	1.66
19	2.99	2.61	2.40	2.27	2.18	2.11	2.06	2.02	1.98	1.96	1.91	1.86	1.81	1.76	1.73	1.70	1.67	1.63
20	2.97	2.59	2.38	2.25	2.16	2.09	2.04	2.00	1.96	1.94	1.89	1.84	1.79	1.74	1.71	1.68	1.64	1.61
21	2.96	2.57	2.36	2.23	2.14	2.08	2.02	1.98	1.95	1.92	1.87	1.83	1.78	1.72	1.69	1.66	1.62	1.59
22	2.95	2.56	2.35	2.22	2.13	2.06	2.01	1.97	1.93	1.90	1.86	1.81	1.76	1.70	1.67	1.64	1.60	1.57
23	2.94	2.55	2.34	2.21	2.11	2.05	1.99	1.95	1.92	1.89	1.84	1.80	1.74	1.69	1.66	1.62	1.59	1.55
24	2.93	2.54	2.33	2.19	2.10	2.04	1.98	1.94	1.91	1.88	1.83	1.78	1.73	1.67	1.64	1.61	1.57	1.53
30	2.88	2.49	2.28	2.14	2.05	1.98	1.93	1.88	1.85	1.82	1.77	1.72	1.67	1.61	1.57	1.54	1.50	1.46
40	2.84	2.44	2.23	2.09	2.00	1.93	1.87	1.83	1.79	1.76	1.71	1.66	1.61	1.54	1.51	1.47	1.42	1.38
60	2.79	2.39	2.18	2.04	1.95	1.87	1.82	1.77	1.74	1.71	1.66	1.60	1.54	1.48	1.44	1.40	1.35	1.29
120	2.75	2.35	2.13	1.99	1.90	1.82	1.77	1.72	1.68	1.65	1.60	1.55	1.48	1.41	1.37	1.32	1.26	1.19
∞	2.71	2.30	2.08	1.94	1.85	1.77	1.72	1.67	1.63	1.60	1.55	1.49	1.42	1.34	1.30	1.24	1.17	1.00

Adapted from Table 18 of Pearson and Hartley 1966.

Table A.5 ***F* Distribution (Continued)**

***p* = .05 values**

Degrees of freedom for denominator υ_2	Degrees of freedom for numerator υ_1																	
	1	2	3	4	5	6	7	8	9	10	12	15	20	30	40	60	120	∞
1	161.4	199.5	215.7	224.6	230.2	234.0	236.8	238.9	240.5	241.9	243.9	245.9	248.0	250.1	251.1	252.2	253.3	254.3
2	18.51	19.00	19.16	19.25	19.30	19.33	19.35	19.37	19.38	19.40	19.41	19.43	19.45	19.46	19.47	19.48	19.49	19.50
3	10.13	9.55	9.28	9.12	9.01	8.94	8.89	8.85	8.81	8.79	8.74	8.70	8.66	8.62	8.59	8.57	8.55	8.53
4	7.71	6.94	6.59	6.39	6.26	6.16	6.09	6.04	6.00	5.96	5.91	5.86	5.80	5.75	5.72	5.69	5.66	5.63
5	6.61	5.79	5.41	5.19	5.05	4.95	4.88	4.82	4.77	4.74	4.68	4.62	4.56	4.50	4.46	4.43	4.40	4.36
6	5.99	5.14	4.76	4.53	4.39	4.28	4.21	4.15	4.10	4.06	4.00	3.94	3.87	3.81	3.77	3.74	3.70	3.67
7	5.59	4.74	4.35	4.12	3.97	3.87	3.79	3.73	3.68	3.64	3.57	3.51	3.44	3.38	3.34	3.30	3.27	3.23
8	5.32	4.46	4.07	3.84	3.69	3.58	3.50	3.44	3.39	3.35	3.28	3.22	3.15	3.08	3.04	3.01	2.97	2.93
9	5.12	4.26	3.86	3.63	3.48	3.37	3.29	3.23	3.18	3.14	3.07	3.01	2.94	2.86	2.83	2.79	2.75	2.71
10	4.96	4.10	3.71	3.48	3.33	3.22	3.14	3.07	3.02	2.98	2.91	2.85	2.77	2.70	2.66	2.62	2.58	2.54
11	4.84	3.98	3.59	3.36	3.20	3.09	3.01	2.95	2.90	2.85	2.79	2.72	2.65	2.57	2.53	2.49	2.45	2.40
12	4.75	3.89	3.49	3.26	3.11	3.00	2.91	2.85	2.80	2.75	2.69	2.62	2.54	2.47	2.43	2.38	2.34	2.30
13	4.67	3.81	3.41	3.18	3.03	2.92	2.83	2.77	2.71	2.67	2.60	2.53	2.46	2.38	2.34	2.30	2.25	2.21
14	4.60	3.74	3.34	3.11	2.96	2.85	2.76	2.70	2.65	2.60	2.53	2.46	2.39	2.31	2.27	2.22	2.18	2.13
15	4.54	3.68	3.29	3.06	2.90	2.79	2.71	2.64	2.59	2.54	2.48	2.40	2.33	2.25	2.20	2.16	2.11	2.07
16	4.49	3.63	3.24	3.01	2.85	2.74	2.66	2.59	2.54	2.49	2.42	2.35	2.28	2.19	2.15	2.11	2.06	2.01
17	4.45	3.59	3.20	2.96	2.81	2.70	2.61	2.55	2.49	2.45	2.38	2.31	2.23	2.15	2.10	2.06	2.01	1.96
18	4.41	3.55	3.16	2.93	2.77	2.66	2.58	2.51	2.46	2.41	2.34	2.27	2.19	2.11	2.06	2.02	1.97	1.92
19	4.38	3.52	3.13	2.90	2.74	2.63	2.54	2.48	2.42	2.38	2.31	2.23	2.16	2.07	2.03	1.98	1.93	1.88
20	4.35	3.49	3.10	2.87	2.71	2.60	2.51	2.45	2.39	2.35	2.28	2.20	2.12	2.04	1.99	1.95	1.90	1.84
21	4.32	3.47	3.07	2.84	2.68	2.57	2.49	2.42	2.37	2.32	2.25	2.18	2.10	2.01	1.96	1.92	1.87	1.81
22	4.30	3.44	3.05	2.82	2.66	2.55	2.46	2.40	2.34	2.30	2.23	2.15	2.07	1.98	1.94	1.89	1.84	1.78
23	4.28	3.42	3.03	2.80	2.64	2.53	2.44	2.37	2.32	2.27	2.20	2.13	2.05	1.96	1.91	1.86	1.81	1.76
24	4.26	3.40	3.01	2.78	2.62	2.51	2.42	2.36	2.30	2.25	2.18	2.11	2.03	1.94	1.89	1.84	1.79	1.73
30	4.17	3.32	2.92	2.69	2.53	2.42	2.33	2.27	2.21	2.16	2.09	2.01	1.93	1.84	1.79	1.74	1.68	1.62
40	4.08	3.23	2.84	2.61	2.45	2.34	2.25	2.18	2.12	2.08	2.00	1.92	1.84	1.74	1.69	1.64	1.58	1.51
60	4.00	3.15	2.76	2.53	2.37	2.25	2.17	2.10	2.04	1.99	1.92	1.84	1.75	1.65	1.59	1.53	1.47	1.39
120	3.92	3.07	2.68	2.45	2.29	2.17	2.09	2.02	1.96	1.91	1.83	1.75	1.66	1.55	1.50	1.43	1.35	1.25
∞	3.84	3.00	2.60	2.37	2.21	2.10	2.01	1.94	1.88	1.83	1.75	1.67	1.57	1.46	1.39	1.32	1.22	1.00

Table A.5 *F* Distribution (Continued)

p = .01 Values

Degrees of freedom for denominator	Degrees of freedom for numerator, υ_1																	
	1	2	3	4	5	6	7	8	9	10	12	15	20	30	40	60	120	∞
1	4052	4999.5	5403	5625	5764	5859	5928	5981	6022	6056	6106	6157	6209	6261	6287	6313	6339	6366
2	98.50	99.00	99.17	99.25	99.30	99.33	99.36	99.37	99.39	99.40	99.42	99.43	99.45	99.47	99.47	99.48	99.49	99.50
3	34.12	30.82	29.46	28.71	28.24	27.91	27.67	27.49	27.35	27.23	27.05	26.87	26.69	26.50	26.41	26.32	26.22	26.13
4	21.20	18.00	16.69	15.98	15.52	15.21	14.98	14.80	14.66	14.55	14.37	14.20	14.02	13.84	13.75	13.65	13.56	13.6
5	16.26	13.27	12.06	11.39	10.97	10.67	10.46	10.29	10.16	10.05	9.89	9.72	9.55	9.38	9.29	9.20	9.11	9.02
6	13.75	10.92	9.78	9.15	8.75	8.47	8.26	8.10	7.98	7.87	7.72	7.56	7.40	7.23	7.14	7.06	6.97	6.88
7	12.25	9.55	8.45	7.85	7.46	7.19	6.99	6.84	6.72	6.62	6.47	6.31	6.16	5.99	5.91	5.82	5.74	5.65
8	11.26	8.65	7.59	7.01	6.63	6.37	6.18	6.03	5.91	5.81	5.67	5.52	5.36	5.20	5.12	5.03	4.95	4.86
9	10.56	8.02	6.99	6.42	6.06	5.80	5.61	5.47	5.35	5.26	5.11	4.96	4.81	4.65	4.57	4.48	4.40	4.31
10	10.04	7.56	6.55	5.99	5.64	5.39	5.20	5.06	4.94	4.85	4.71	4.56	4.41	4.25	4.17	4.08	4.00	3.91
11	9.65	7.21	6.22	5.67	5.32	5.07	4.89	4.74	4.63	4.54	4.40	4.25	4.10	3.94	3.86	3.78	3.69	3.60
12	9.33	6.93	5.95	5.41	5.06	4.82	4.64	4.50	4.39	4.30	4.16	4.01	3.86	3.70	3.62	3.54	3.45	3.36
13	9.07	6.70	5.74	5.21	4.86	4.62	4.44	4.30	4.19	4.10	3.96	3.82	3.66	3.51	3.43	3.34	3.25	3.17
14	8.86	6.51	5.56	5.04	4.69	4.46	4.28	4.14	4.03	3.94	3.80	3.66	3.51	3.35	3.27	3.18	3.09	3.00
15	8.68	6.36	5.42	4.89	4.56	4.32	4.14	4.00	3.89	3.80	3.67	3.52	3.37	3.21	3.13	3.05	2.96	2.87
16	8.53	6.23	5.29	4.77	4.44	4.20	4.03	3.89	3.78	3.69	3.55	3.41	3.26	3.10	3.02	2.93	2.84	2.75
17	8.40	6.11	5.18	4.67	4.34	4.10	3.93	3.79	3.68	3.59	3.46	3.31	3.16	3.00	2.92	2.83	2.75	2.65
18	8.29	6.01	5.09	4.58	4.25	4.01	3.84	3.71	3.60	3.51	3.37	3.23	3.08	2.92	2.84	2.75	2.66	2.57
19	8.18	5.93	5.01	4.50	4.17	3.94	3.77	3.63	3.52	3.43	3.30	3.15	3.00	2.84	2.76	2.67	2.58	2.49
20	8.10	5.85	4.94	4.43	4.10	3.87	3.70	3.56	3.46	3.37	3.23	3.09	2.94	2.78	2.69	2.61	2.52	2.42
21	8.02	5.78	4.87	4.37	4.04	3.81	3.64	3.51	3.40	3.31	3.17	3.03	2.88	2.72	2.64	2.55	2.46	2.36
22	7.95	5.72	4.82	4.31	3.99	3.76	3.59	3.45	3.35	3.26	3.12	2.98	2.83	2.67	2.58	2.50	2.40	2.31
23	7.88	5.66	4.76	4.26	3.94	3.71	3.54	3.41	3.30	3.21	3.07	2.93	2.78	2.62	2.54	2.45	2.35	2.26
24	7.82	5.61	4.72	4.22	3.90	3.67	3.50	3.36	3.26	3.17	3.03	2.89	2.74	2.58	2.49	2.40	2.31	2.21
30	7.56	5.39	4.51	4.02	3.70	3.47	3.30	3.17	3.07	2.98	2.84	2.70	2.55	2.39	2.30	2.21	2.11	2.01
40	7.31	5.18	4.31	3.83	3.51	3.29	3.12	2.99	2.89	2.80	2.66	2.52	2.37	2.20	2.11	2.02	1.92	1.80
60	7.08	4.98	4.13	3.65	3.34	3.12	2.95	2.82	2.72	2.63	2.50	2.35	2.20	2.03	1.94	1.84	1.73	1.60
120	6.85	4.79	3.95	3.48	3.17	2.96	2.79	2.66	2.56	2.47	2.34	2.19	2.03	1.86	1.76	1.66	1.53	1.38
∞	6.63	4.61	3.78	3.32	3.02	2.80	2.64	2.51	2.41	2.32	2.18	2.04	1.88	1.70	1.59	1.47	1.32	1.00

Table A.6 χ^2 **Distribution**

For various degrees of freedom df, the tabled entries represent the values of χ^2 above which a proportion p of the distribution falls. (Example: for df = 5, a $\chi^2 = 11.070$ is exceeded by $p = .05$ or 5% of the distribution.)

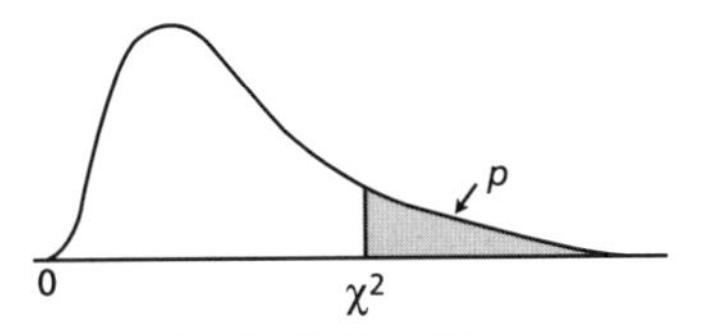

	p						
df	.99	.95	.90	.10	.05	.01	.001
1	$.0^{3}157$	.00393	.0158	2.706	3.841	6.635	10.827
2	.0201	.103	.211	4.605	5.991	9.210	13.815
3	.115	.352	.584	6.251	7.815	11.345	16.266
4	.297	.711	1.064	7.779	9.488	13.277	18.467
5	.554	1.145	1.610	9.236	11.070	15.086	20.515
6	.872	1.635	2.204	10.645	12.592	16.812	22.457
7	1.239	2.167	2.833	12.017	14.067	18.475	24.322
8	1.646	2.733	3.490	13.362	15.507	20.090	26.125
9	2.088	3.325	4.168	14.684	16.919	21.666	27.877
10	2.558	3.940	4.865	15.987	18.307	23.209	29.588
11	3.053	4.575	5.578	17.275	19.675	24.725	31.264
12	3.571	5.226	6.304	18.549	21.026	26.217	32.909
13	4.107	5.892	7.042	19.812	22.362	27.688	34.528
14	4.660	6.571	7.790	21.064	23.685	29.141	36.123
15	5.229	7.261	8.547	22.307	24.996	30.578	37.697
16	5.812	7.962	9.312	23.542	26.296	32.000	39.252
17	6.408	8.672	10.085	24.769	27.587	33.409	40.790
18	7.015	9.390	10.865	25.989	28.869	34.805	42.312
19	7.633	10.117	11.651	27.204	30.144	36.191	43.820
20	8.260	10.851	12.443	28.412	31.410	37.566	45.315
21	8.897	11.591	13.240	29.615	32.671	38.932	46.797
22	9.542	12.338	14.041	30.813	33.924	40.289	48.268
23	10.196	13.091	14.848	32.007	35.172	41.638	49.728
24	10.856	13.848	15.659	33.196	36.415	42.980	51.179
25	11.524	14.611	16.473	34.382	37.652	44.314	52.620
26	12.198	15.379	17.292	35.563	38.885	45.642	54.052
27	12.879	16.151	18.114	36.741	40.113	46.963	55.476
28	13.565	16.928	18.939	37.916	41.337	48.278	56.893
29	14.256	17.708	19.768	39.087	42.557	49.588	58.302
30	14.953	18.493	20.599	40.256	43.773	50.892	59.703

Adapted from Table IV of Fisher and Yates (1974).

Table A.7 **Coefficients $\{a_{n-i+1}\}$ for the Shapiro–Wilk W test for normality, for $n = 2(1)50$.**

	n								
i	**2**	**3**	**4**	**5**	**6**	**7**	**8**	**9**	**10**
1	0.7071	0.7071	0.6872	0.6646	0.6431	0.6233	0.6052	0.5888	0.5739
2	—	.0000	.1677	.2413	.2806	.3031	.3164	.3244	.3291
3	—	—	—	.0000	.0875	.1401	.1743	.1976	.2141
4	—	—	—	—	—	.0000	.0561	.0947	.1224
5	—	—	—	—	—	—	—	.0000	.0399

	n									
i	**11**	**12**	**13**	**14**	**15**	**16**	**17**	**18**	**19**	**20**
1	0.5601	0.5475	0.5359	0.5251	0.5150	0.5056	0.4968	0.4886	0.4808	0.4734
2	.3315	.3325	.3325	.3318	.3306	.3290	.3273	.3253	.3232	.3211
3	.2260	.2347	.2412	.2460	.2495	.2521	.2540	.2553	.2561	.2565
4	.1429	.1586	.1707	.1802	.1878	.1939	.1988	.2027	.2059	2085
5	.0695	.0922	.1099	.1240	.1353	.1447	.1524	.1587	.1641	.1686
6	0.0000	0.0303	0.0539	0.0727	0.0880	0.1005	0.1109	0.1197	0.1271	0.1334
7	—	—	.0000	.0240	.0433	.0593	.0725	.0837	.0932	.1013
8	—	—	—	—	.0000	.0196	.0359	.0496	.0612	.0711
9	—	—	—	—	—	—	.0000	.0163	.0303	.0422
10	—	—	—	—	—	—	—	—	.0000	.0140

	n									
i	**21**	**22**	**23**	**24**	**25**	**26**	**27**	**28**	**29**	**30**
1	0.4643	0.4590	0.4542	0.4493	0.4450	0.4407	0.4366	0.4328	0.4291	0.4254
2	.3185	.3156	.3126	.3098	.3069	.3043	.3018	.2992	.2968	.2944
3	.2578	.2571	.2563	.2554	.2543	.2533	.2522	.2510	.2499	.2487
4	.2119	.2131	.2139	.2145	.2148	.2151	.2152	.2151	.2150	.2148
5	.1736	.1764	.1787	.1807	.1822	.1836	.1848	.1857	.1864	.1870
6	0.1399	0.1443	0.1480	0.1512	0.1539	0.1563	0.1584	0.1601	0.1616	0.1630
7	.1092	.1150	.1201	.1245	.1283	.1316	.1346	.1372	.1395	.1415
8	.0804	.0878	.0941	.0997	.1046	.1089	.1128	1162	1192	.1219
9	.0530	.0618	.0696	.0764	.0823	.0876	.0923	.0965	.1002	.1036
10	.0263	.0368	.0459	.0539	.0610	.0672	.0728	.0778	.0822	.0862
11	0.0000	0.0122	0.0228	0.0321	0.0403	0.0476	0.0540	0.0598	0.0650	0.0697
12	—	—	.0000	.0107	.0200	.0284	.0358	.0424	.0483	.0537
13	—	—	—	—	.0000	.0094	.0178	.0253	.0320	.0381
14	—	—	—	—	—	—	.0000	.0084	.0159	.0227
15	—	—	—	—	—	—	—	—	.0000	.0076

Source: Wetherill (1981).

Note: The notation $\mathbf{n} = 2(1)50$ means that entries are provided for $\mathbf{n} = 2$ to $\mathbf{n} = 50$ with increments of 1.

Table A.7 **(Continued)**

i	*n* 31	32	33	34	35	36	37	38	39	40
1	0.4220	0.4188	0.4156	0.4127	0.4096	0.4068	0.4040	0.4015	0.3989	0.3964
2	.2921	.2898	.2876	.2854	.2834	.2813	.2794	.2774	.2755	.2737
3	.2475	.2463	.2451	.2439	.2427	.2415	.2403	.2391	.2380	.2368
4	.2145	.2141	.2137	.2132	.2127	.2121	.2116	.2110	.2104	.2098
5	.1874	.1878	.1880	.1882	.1883	.1883	.1883	.1881	.1880	.1878
6	0.1641	0.1651	0.1660	0.1667	0.1673	0.1678	0.1683	0.1686	0.1689	0.1691
7	.1433	.1449	.1463	.1475	.1487	.1496	.1505	.1513	.1520	.1526
8	.1243	.1265	.1284	.1301	.1317	.1331	.1344	.1356	.1366	.1376
9	.1066	.1093	.1118	.1140	.1160	.1179	.1196	.1211	.1225	.1237
10	.0899	.0931	.0961	.0988	.1013	.1036	.1056	.1075	.1092	.1108
11	0.0739	0.0777	0.0812	0.0844	0.0873	0.0900	0.0924	0.0947	0.0967	0.0986
12	.0585	.0629	.0669	.0706	.0739	.0770	.0798	.0824	.0848	.0870
13	.0435	.0485	.0530	.0572	.0610	.0645	.0677	.0706	.0733	.0759
14	.0289	.0344	.0395	.0441	.0484	.0523	.0559	.0592	.0622	.0651
15	.0144	.0206	.0262	.0314	.0361	.0404	.0444	.0481	.0515	.0546
16	0.0000	0.0068	0.0131	0.0187	0.0239	0.0287	0.0331	0.0372	0.0409	0.0444
17	—	—	.0000	.0062	.0119	.0172	.0220	.0264	.0305	.0343
18	—	—	—	—	.0000	.0057	.0110	.0158	.0203	.0244
19	—	—	—	—	—	—	.0000	.0053	.0101	.0146
20	—	—	—	—	—	—	—	—	.0000	.0049

i	*n* 41	42	43	44	45	46	47	48	49	50
1	0.3940	0.3917	0.3894	0.3872	0.3850	0.3830	0.3808	0.3789	0.3770	0.3751
2	.2719	.2701	.2684	.2667	.2651	.2635	.2620	.2604	.2589	.2574
3	.2357	.2345	.2334	.2323	.2313	.2302	.2291	.2281	.2271	.2260
4	.2091	.2085	.2078	.2072	.2065	.2058	.2052	.2045	.2038	.2032
5	.1876	.1874	.1871	.1868	.1865	.1862	.1859	.1855	.1851	.1847
6	0.1693	0.1694	0.1695	0.1695	0.1695	0.1695	0.1695	0.1693	0.1692	0.1691
7	.1531	.1535	.1539	.1542	.1545	.1548	.1550	.1551	.1553	.1554
8	.1384	.1392	.1398	.1405	.1410	.1415	.1420	.1423	.1427	.1430
9	.1249	.1259	.1269	.1278	.1286	.1293	.1300	.1306	.1312	.1317
10	.1123	.1136	.1149	.1160	.1170	.1180	.1189	.1197	.1205	.1212
11	0.1004	0.1020	0.1035	0.1049	0.1062	0.1073	0.1085	0.1095	0.1105	0.1113
12	.0891	.0909	.0927	.0943	.0959	.0972	.0986	.0998	.1010	.1020
13	.0782	.0804	.0824	.0842	.0860	.0876	.0892	.0906	.0919	.0932
14	.0677	.0701	.0724	.0745	.0765	.0783	.0801	.0817	.0832	.0846
15	.0575	.0602	.0628	.0651	.0673	.0694	.0713	.0731	.0748	.0764
16	0.0476	0.0506	0.0534	0.0560	0.0584	0.0607	0.0628	0.0648	0.0667	0.0685
17	.0379	.0411	.0442	.0471	.0497	.0522	.0546	.0568	.0588	.0608
18	.0283	.0318	.0352	.0383	.0412	.0439	.0465	.0489	.0511	.0532
19	.0188	.0227	.0263	.0296	.0328	.0357	.0385	.0411	.0436	.0459
20	.0094	0136	.0175	.0211	.0245	.0277	.0307	.0335	.0361	.0386
21	0.0000	0.0045	0.0087	0.0126	0.0163	0.0197	0.0229	0.0259	0.0288	0.0314
22	—	—	.0000	.0042	.0081	.0118	.0153	.0185	.0215	.0244
23	—	—	—	—	.0000	.0039	.0076	.0111	.0143	.0174
24	—	—	—	—	—	—	.0000	.0037	.0071	.0104
25	—	—	—	—	—	—	—	—	.0000	.0035

Table A.8 **Percentage points of the *W* test* for *n* = 3(1)50**

n	Level								
	0.01	**0.02**	**0.05**	**0.10**	**0.50**	**0.90**	**0.95**	**0.98**	**0.99**
3	0.753	0.756	0.767	0.789	0.959	0.998	0.999	1.000	1.000
4	.687	.707	.748	.792	.935	.987	.992	.996	.997
5	.686	.715	.762	.806	.927	.979	.986	.991	.993
6	0.713	0.743	0.788	0.826	0.927	0.974	0.981	0.986	0.989
7	.730	.760	.803	.838	.928	.972	.979	.985	.988
8	.749	.778	.818	.851	.932	.972	.978	.984	.987
9	.764	.791	.829	.859	.935	.972	.978	.984	.986
10	.781	.806	.842	.869	.938	.972	.978	.983	.986
11	0.792	0.817	0.850	0.876	0.940	0.973	0.979	0.984	0.986
12	.805	.828	.859	.883	.943	.973	.979	.984	.986
13	.814	.837	.866	.889	.945	.974	.979	.984	.986
14	.825	.846	.874	895	.947	.975	.980	.984	.986
15	.835	.855	.881	.901	.950	.975	.980	.984	.987
16	0.844	0.863	0.887	0.906	0.952	0.976	0.981	0.985	0.987
17	.851	.869	.892	.910	.954	.977	.981	.985	.987
18	.858	.874	.897	.914	.956	.978	.982	.986	.988
19	.863	.879	.901	.917	.957	.978	.982	.986	.988
20	.868	.884	.905	.920	.959	.979	.983	.986	.988
21	0.873	0.888	0.908	0.923	0.960	0.980	0.983	0.987	0.989
22	.878	.892	.911	.926	.961	.980	.984	.987	.989
23	.881	.895	.914	.928	.962	.981	.984	.987	.989
24	.884	.898	.916	.930	.963	.981	.984	.987	.989
25	.888	.901	.918	.931	.964	.981	0985	.988	.989
26	0.891	0.904	0.920	0.933	0.965	0.982	0.985	0.988	0.989
27	.894	.906	.923	.935	.965	.982	.985	.988	.990
28	.896	.908	.924	.936	.966	.982	.985	.988	.990
29	.898	.910	.926	.937	.966	.982	.985	.988	.990
30	.900	.912	.927	.939	.967	.983	.985	.988	.900
31	0.902	0.914	0.929	0.940	0.967	0.983	0.986	0.988	0.990
32	.904	.915	.930	.941	.968	.983	.986	.988	.990
33	.906	.917	.931	.942	.968	.983	.986	.989	.990
34	.908	.919	.933	.943	.969	.983	.986	.989	.990
35	.910	.920	.934	.944	.969	.984	.986	.989	.990
36	0.912	0.922	0.935	0.945	0.970	0.984	0.986	0..989	0.990
37	.914	.924	.936	.946	.970	.984	.987	.989	.990
38	.916	.925	.938	.947	.971	.984	.987	.989	.990
39	.917	.927	.939	.948	.971	.984	.987	.989	.991
40	.919	.928	.940	.949	.972	.985	.987	.989	.991
41	0.920	0.929	0.941	0.950	0.972	0.985	0.987	0.989	0.991
42	.922	.930	.942	.951	.972	.985	.987	.989	.991
43	.923	.932	.943	.951	.973	.985	.987	.990	.991
44	.924	.933	.944	.952	.973	.985	.987	.990	.991
45	.926	.934	.945	.953	.973	.985	.988	.990	.991
46	0.927	0.935	0.945	0.953	0.974	0.985	0.988	0.990	0.991
47	.928	.936	.946	.954	.974	.985	.988	.990	.991
48	.929	.937	.947	.954	.974	.985	.988	.990	.991
49	.929	.937	.947	.955	.974	.985	.988	.990	.991
50	.930	.938	.947	.955	.974	.985	.988	.990	.991

*Based on fitted Johnson (1949) S_B approximation; see Shapiro and Wilk (1965) for details.

Source: Wetherill (1981).

Note: The notation **n** = 3(1)50 means that entries are provided for **n** = 3 to **n** = 50 with increments of 1.

Appendix B: Review and Extension of Some Probability Theory

A discrete random variable, X, has a probability distribution (sometimes called a probability mass function) denoted by $P(X=x)=p(x)$, where x is the value taken by X. A continuous random variable has a probability distribution (also called a probability density function or pdf) denoted by $f(x)$. The likelihood of getting a specific value x is zero, since the distribution is continuous. The likelihood of getting a value within a range, $a<x<b$ is equal to the area under the curve, $f(x)$, that lies between a and b. For those familiar with calculus, this area is given as the integral of $f(x)$ from a to b:

$$P(X = a < x < b) = \int_a^b f(x)\,\mathrm{d}x \tag{B.1}$$

For those not familiar with calculus, the integral sign may be thought of as similar to the summation sign; the only difference is that with a continuous random variable, we have an infinite number of values to sum over. Since the probability of getting a value between minus and plus infinity is equal to one, the total area under the curve $f(x)$ must equal one:

$$\int_{-\infty}^{+\infty} f(x)\,\mathrm{d}x = 1 \tag{B.2}$$

Cumulative distribution functions tell us the likelihood that the random variable will be less than or equal to a particular value. For a discrete random variable, the probability of obtaining a value less than or equal to a is

$$F(a) = \sum_{x \le a} p(x) \tag{B.3}$$

For a continuous random variable, we have

$$F(a) = p(X \le a) = \int_{-\infty}^{a} f(x) \tag{B.4}$$

Expected Values

The expected value of a random variable, $E[X]$, is also known as the theoretical mean and is denoted by μ. The expected value is given as the weighted average of the possible values the random variable can take on, where the weights are the likelihoods of obtaining those values. For a discrete random variable,

$$E[X] = \mu = \sum_x \frac{xp(x)}{\sum_x p(x)} = \sum_x xp(x) \tag{B.5}$$

For a continuous random variable,

$$E[X] = \mu = \frac{\int_{-\infty}^{+\infty} xf(x)\,\mathrm{d}x}{\int_{-\infty}^{+\infty} f(x)\,\mathrm{d}x} = \int_{-\infty}^{+\infty} x\,f(x)\,\mathrm{d}x \qquad \text{(B.6)}$$

As an example, consider the experiment that consists of tossing a die. What is the expected value of the die? This is equivalent to asking what the expected average is of a large number of tosses. Using Equation B.5,

$$\sum_{x=1}^{6} 1\left(\frac{1}{6}\right) + 2\left(\frac{1}{6}\right) + 3\left(\frac{1}{6}\right) + \cdots + 6\left(\frac{1}{6}\right) = 3.5 \qquad \text{(B.7)}$$

What is the expected value of a uniform random variable – that is, a random variable that has equally likely outcomes over the range (a, b)? Such a random variable has a probability density function given by

$$f(x) = \frac{1}{b-a} \qquad \text{(B.8)}$$

The expected value is then

$$E[X] = \mu = \int_{a}^{b} \frac{x}{b-a}\,\mathrm{d}x = \frac{1}{b-a}\left(\frac{b^2-a^2}{2}\right) = \frac{a+b}{2} \qquad \text{(B.9)}$$

The expected value of any function of a random variable, $g(x)$, is a weighted average of the values of $g(x)$, where the weights are again the likelihoods of obtaining the values of $g(x)$. Thus we have

$$\left.\begin{aligned} E[g(X)] &= \sum_{x} g(x)p(x) \\ E[g(X)] &= \int_{-\infty}^{+\infty} g(x)f(x)\,\mathrm{d}x \end{aligned}\right\} \qquad \text{(B.10)}$$

for discrete and continuous random variables, respectively. Useful rules for working with expected values are: (i) the expected value of a constant is simply the constant; (ii) the expected value of a constant times a random variable is the constant times the expected value of the random variable; and (iii) the expected value of a sum is equal to the sum of the expected values. These rules are summarized below:

$$\left.\begin{aligned} E[a] &= a && \text{(i)} \\ E[bX] &= bE[X] && \text{(ii)} \\ E[a+bX] &= a+bE[X] && \text{(i, ii, and iii)} \end{aligned}\right\} \qquad \text{(B.11)}$$

Variance of a Random Variable

The variance of a random variable, $\sigma^2 = V[X]$, is the expected value of the squared deviation of an observation from the mean:

$$V[X] = \sigma^2 = E[(X - E[X])^2 = E[(X - \mu)^2] \tag{B.12}$$

Using the rules for expected values above,

$$V[X] = E[X^2 - 2X\mu + \mu^2] = E[X^2] - 2\mu E[X] + \mu^2 = E[X^2] - \mu^2 \tag{B.13}$$

To illustrate, let us return to the experiment involving the toss of a die. The variance of the random variable X in this case is equal to $E[X^2] - 3.5^2$. The expected value of X^2 is found using Equation B.10:

$$E[X^2] = 1^2\left(\frac{1}{6}\right) + 2^2\left(\frac{1}{6}\right) + \cdots + 6^2\left(\frac{1}{6}\right) = 15.17 \tag{B.14}$$

The variance is therefore equal to $15.17 - 3.5^2 = 2.92$. To illustrate the derivation of the variance using a continuous variable, let us continue with the example of a uniform random variable. We have

$$E[X^2] \int_a^b x^2 \frac{1}{b-a} = \frac{b^3 - a^3}{3(b-a)} \tag{B.15}$$

Then

$$V[X] = \frac{b^3 - a^3}{3(b-a)} - \frac{(a+b)^2}{4} = \frac{b-a}{12} \tag{B.16}$$

Covariance of Random Variables

How do two variables co-vary? Is there a tendency for one variable to exhibit high variables when the other does? Or are the variables independent? The covariance of two random variables, X and Y, is defined as the expected value of the product of the two deviations from the means:

$$\text{Cov}[X, Y] = E[(X - \mu_X)(Y - \mu_Y)] \tag{B.17}$$

This may be rewritten in the form

$$\text{Cov}[X, Y] = E[(XY - X\mu_Y - Y\mu_X + \mu_X\mu_Y)] = E[XY] - \mu_X\,\mu_y \tag{B.18}$$

To find the observed covariance for a set of data, we find the average value of the product of deviations:

$$\text{Cov}[x, y] = \sum_{i=1}^{n} \frac{(x_i - \bar{x})(y_i - \bar{y})}{n} \tag{B.19}$$

The correlation coefficient is the standardized covariance:

$$\rho = \frac{\text{Cov}[X, Y]}{\sigma_X \sigma_Y} \tag{B.20}$$

Bibliography

Ackerman, W.V. 1998. Socioeconomic correlates of increasing crime rates in smaller communities. *Professional Geographer* 50: 372–87.

Allen, J.P. and Turner, E. 1996. Spatial patterns of immigrant assimilation. *Professional Geographer* 48: 140–55.

Andrews, D.F. 1985. *Data: a collection of problems from many fields for the student and research worker*. New York: Springer-Verlag.

Anselin, L. 1992. *SpaceStat*: A program for the analysis of spatial data. Santa Barbara, CA: National Center for Geographic Information and Analysis.

Anselin, L. 1995. Local indicators of spatial association – LISA. *Geographical Analysis* 27: 93–115.

Bailey, A. and Gatrell, A. 1995. *Interactive spatial data analysis*. Essex: Longman (published in the U.S. by Wiley).

Besag, J. and Newell, J. 1991. The detection of clusters in rare diseases. *Journal of the Royal Statistical Society, Series A* 154: 143–55.

Brunsdon, C., Fotheringham, A.S., and Charlton, M. 1996. Geographically weighted regression: a method for exploring spatial nonstationarity. *Geographical Analysis* 28: 281–98.

Brunsdon, C., Fotheringham, A.S., and Charlton, M. 1999. Some notes on parametric significance tests for geographically weighted regression. *Journal of Regional Science* 39: 497–524.

Casetti, E. 1972. Generating models by the expansion method: applications to geographic research. *Geographical Analysis* 4: 81–91.

Clark, P.J. and Evans, F.C. 1954. Distance to nearest neighbor as a measure of spatial relationships in populations. *Ecology* 35: 445–53.

Clarke, A.E. and Holly, B. 1996. The organization of production in high technology industries: an empirical assessment. *Professional Geographer* 48: 127–39.

Cliff, A. and Ord, J.K. 1975. The comparison of means when samples consist of spatial autocorrelated observations. *Environment and Planning A* 7: 725–734.

Clifford, P. and Richardson, S. 1985. Testing the association between two spatial processes. *Statistics and Decisions* Supplement Number 2, 155–60.

Cohen, J.E. 1995. How many people can the earth support? New York: W.W. Norton and Co.

Comrie, A.C. 1996. An all-season synoptic climatology of air pollution in the U.S. Mexico border region. *Professional Geogapher* 48: 237–51.

Cornish, S.L. 1997. Strategies for the acquisition of market intelligence and implications for the transferability of information inputs. *Annals of the Association of American Geographers* 87: 451–70.

Cressie, N. 1993. *Statistical analysis of spatial data*. New York: Wiley.

Curtiss, J. and McIntosh, R. 1950. The interrelations of certain analytic and synthetic phytosociological characters. *Ecology* 31: 434–55.

Dagel, K.C. 1997. Defining drought in marginal areas: the role of perception. *Professional Geographer* 49: 192–202.

Easterlin, R. 1980. *Birth and fortune: the impact of numbers on personal welfare*. New York: Basic Books.

Fan, C. 1996. Economic opportunities and internal migration: a case study of Guangdong Province, China. *Professional Geographer* 48: 28–45.

Fisher, R.A., and Yates, F. 1974. *Statistical Tables for Biological, Agricultural, and Medical Research*, 6th edition. London: Longman.

Fotheringham, A.S. and Rogerson, P. 1993. GIS and spatial analytical problems. *International Journal of Geographical Information Systems* 7: 3–19.

Fotheringham, A.S. and Wong, D. 1991. The modifiable area unit problem in multivariate statistical analysis. *Environment and Planning A* 23: 1025–44.

Fotheringham, A.S., Charlton, M.E., and Brunsdon, C. 1998. Geographically weighted regression: a natural evolution of the expansion method for spatial data analysis. *Environment and Planning A* 30: 1905–27.

Fotheringham, A.S., Brunsdon, C., and Charlton, M. 2000. *Quantitative geography: perspectives on spatial data analysis*. London: Sage Publications.

Gehlke, C. and Biehl, K. 1934. Certain effects of grouping upon the size of the correlation coefficient in census tract material. *Journal of the American Statistical Association* 29: 169–70.

Gott, R. 1993. Implications of the Copernican principle for our future prospects. *Nature* 363: 315–19.

Griffith, D.A. 1978. A spatially adjusted ANOVA model. *Geographical Analysis* 10: 296–301.

Griffith, D.A. 1987. *Spatial autocorrelation: a primer*. Washington, DC: Association of American Geographers.

Griffith, D.A. 1996. Computational simplifications for space–time forecasting within GIS: the neighbourhood spatial forecasting model. In *Spatial analysis: modelling in a GIS environment* (Eds. P. Longley and M. Batty), pp. 247–60. Cambridge: Geoinformation International (distributed by Wiley).

Griffith, D.A., Doyle, P.G., Wheeler, D.C., and Johnson, D.L. 1998. A tale of two swaths: urban childhood blood-lead levels across Syracuse, New York. *Annals of the Association of American Geographers* 88: 640–65.

Hadi, A.S. and Ling, R.F. 1998. Some cautionary notes on the use of principal components regression. *American Statistician* 52,1: 15–19.

Haining, R. 1990a. *Spatial data analysis in the social and environmental sciences*. Cambridge: Cambridge University Press.

Haining, R. 1990b. The use of added variable plots in regression modelling with spatial data. *The Professional Geographer* 42: 336–45.

Johnson, B.W. and McCulloch, R.E. 1987. Added variable plots in linear regression. *Technometrics* 29: 427–33.

Jones III, J.P. and Casetti, E. 1992. *Applications of the expansion method*. London: Routledge.

Keim, B.D. 1997. Preliminary analysis of the temporal patterns of heavy rainfall across the Southeastern United States. *Professional Geographer* 49: 94–104.

Longley, P., Brooks, S.M., McDonnell, R., and Macmillan, B. 1998. *Geocomputation: A primer*. Chichester: Wiley.

MacDonald, G.M., Szeicz, J.M., Claricoates, J., and Dale, K.A. 1998. Response of the central Canadian treeline to recent climatic changes. *Annals of the Association of American Geographers* 88: 183–208.

Mallows, C. 1998. The zeroth problem. *American Statistician* 52,1: 1–9.

Meehl, P. 1990. Why summaries of research on psychological theories are often uninterpretable. *Psychological Reports* 66: 195–244 (Monograph Supplement 1-V66).
Moran, P.A.P. 1948. The interpretation of statistical maps. *Journal of the Royal Statistical Society, Series B* 10: 245–51.
Moran, P.A.P. 1950. Notes on continuous stochastic phenomena. *Biometrika* 37:17–23.
Murtagh, F. 1985. A survey of algorithms for contiguity-constrained clustering and related problems. *The Computer Journal* 28: 82–88.
Myers, D., Lee, S.W., and Choi, S.S. 1997. Constraints of housing age and migration on residential mobility. *Professional Geographer* 49: 14–28.
Nelson, P.W. 1997. Migration, sources of income, and community change in the Pacific Northwest. *Professional Geographer* 49: 418–30.
O'Loughlin, J., Ward, M.D., Lofdahl, C.L., Cohen, J.S., Brown, D.S., Reilly, D., Gleditsch, K.S., and Shin, M. 1998. The diffusion of democracy, 1946–1994. *Annals of the Association of American Geographers* 88: 545–74.
Ord, J.K. and Getis, A. 1995. Local spatial autocorrelation statistics: distributional issues and an application. *Geographical Analysis* 27: 286–306.
O'Reilly, K. and Webster, G.R. 1998. A sociodemographic and partisan analysis of voting in three anti-gay rights referenda in Oregon. *Professional Geographer* 50: 498–515.
Ormrod, R.K. and Cole, D.B. 1996. The vote on Colorado's Amendment Two. *Professional Geographer* 48: 14–27.
Pearson, E.S. and Hartley, H.O. (Eds.) 1966. *Biometrika Tables for Statisticians*, Vol. 1. Cambridge: Cambridge University Press.
Plane, D. and Rogerson, P. 1991. Tracking the baby boom, the baby bust, and the echo generations: how age composition regulates US migration. *Professional Geographer* 43: 416–39.
Plane, D. and Rogerson, P. 1994. *The geographical analysis of population: with applications to planning and business*. New York: Wiley.
Robinson, W. 1950. Ecological correlation and the behavior of individuals. *American Sociological Review* 15: 351–57.
Rogers, A. 1975. *Matrix population models*. Thousand Oaks, CA: Sage Publications.
Rogerson, P. 1987. Changes in U.S. national mobility levels. *Professional Geographer* 39: 344–51.
Rogerson, P. and Plane, D. 1998. The dynamics of neighborhood age composition. *Environment and Planning A* 30: 1461–72.
Sachs, L. 1984. *Applied statistics: a handbook of techniques*. New York: Springer-Verlag.
Scheffé, H. 1959. *The analysis of variance*. New York: Wiley.
Shapiro, S.S. and Wilk, M.B. 1965. An analysis of variance test for normality. *Biometrika* 52: 591–611.
Slocum, T. 1990. The use of quantitative methods in major geographical journals, 1956–1986. *Professional Geographer* 42: 84–94.
Standing, L., Sproule, R., and Khouzam, N. 1991. Empirical statistics: IV. Illustrating Meehl's sixth law of soft psychology: everything correlates with everything. *Psychological Reports* 69: 123–26.
Stouffer, S. 1940. Intervening opportunities: a theory relating mobility and distance. *American Sociological Review* 5: 845–67.
Tukey, J. W. 1972. Some graphic and semigraphic displays. In *Statistical papers in honor of George W. Snedecor* (Ed. T.A. Bancroft). Lowa State University Press. Ames, IA.
Velleman, P.F. and Hoaglin, D.C. 1981. *Applications, basics, and computing of exploratory data analysis*. Belmont, CA: Wadsworth.
Webster, G.R. 1996. Partisan shifts in presidential and gubernatorial elections in Alabama, 1932–94. *Professional Geographer* 48: 379–91.

Weisberg, S. 1985. *Applied linear regression*. New York: Wiley.
Wetherill, G.B. 1981. *Intermediate statistical methods*. London: Chapman Hall.
Williams, K.R.S. and Parker, K.C. 1997. Trends in interdiurnal temperature variation for the Central United States, 1945–1985. *Professional Geographer* 49: 342–355.
Wyllie, D.S. and Smith, G.C. 1996. Effects of extroversion on the routine spatial behavior of middle adolescents. *Professional Geographer* 48: 166–80.

Index